Do you trust God enough to pray about everything?

COSMIC CONVERSATIONS WITH OUR CREATOR

Enhance Your Prayer Life by Embracing His Power as the Creator of All Things, Great and Small

EUGENE B. (CHIP) AYDELOTT, M.DIV.

Cosmic Conversations with Our Creator: Enhance Your Prayer Life by Embracing His Power as the Creator of All Things, Great and Small

ISBN: 979-8-9948217-0-1

Printed in the United States of America.

PRAISE FOR *COSMIC CONVERSATIONS WITH OUR CREATOR*

"Likewise the Spirit helps us in our weakness. For we do not know what to pray for as we ought, but the Spirit himself intercedes for us with groanings too deep for words. And he who searches hearts knows what is the mind of the Spirit, because the Spirit intercedes for the saints according to the will of God."

—Romans 8:26–27

"According to Romans 8:26–27, when you do not know what to pray for, the Holy Spirit intercedes by praying on our behalf. These are very comforting words for those moments or periods in our lives when we are truly without words. At other times, we may wish to formulate our own prayers. Chip helps us put words to them. From the simplest prayer, "God, please help me" to heartfelt personal examples from his own life, this book will help you draw closer to God and develop your own personal prayer life."

Gordon Linam, Aquatic Biologist
MS, Fisheries and Wildlife Science

"As a Christian who has spent years studying and teaching the sciences (chemistry and physics, primarily), I have had the struggle of reconciling my belief in a Divine Creator with what science has taught us about how nature adjusts to change through the process of evolution. Chip recounts how his education in biology and theology led him to see that 'science and faith complement one another.' He describes how the complexity of both the microscopic and macroscopic world points to a Creator and reminds us that He, the Creator of all, wants to be in communion with us. Through his personal story of heartache, loss, and a tumor named Roger, Chip encourages us to harness God's love and power through the simple act of prayer."

Michele Lumpkin, BS, Chemistry; MAT, Chemistry
Retired Chemistry and Physics Teacher

"Inspiring and informative, this book shows that theology and science are not enemies but complementary ways of awakening awe and encountering God. *Cosmic Conversations with Our Creator* offers practical guidance, surprising scientific vignettes, and deeply pastoral wisdom that renewed my awe for God's creation and His love for me. This book encouraged me to get closer to God through prayer."

Kim Ricca
Christ Follower, Wife, Mom

TABLE OF CONTENTS

ACKNOWLEDGMENTS

A special thank you to my lifelong friend, Libbye Morris, without whose patience, encouragement, and literary expertise this book would never have come into existence.

Thank you to the people who reviewed this book and provided feedback, including Gordon and Leanne Lynam, Michelle Lumpkin, and Kim Ricca.

I'd also like to thank the entire staff at Neurosurgery One and CommonSpirit, St. Anthony Hospital in Lakewood, Colorado, both for their medical expertise and for their compassionate care in dealing with my brain tumor, "Roger."

Also, I want to thank my kids—Rob and Taylor, and Lexie and Arron. I love you all very much. You guys are the best. I know my health issues and my mistakes have complicated your lives, and I thank you so much for your patience and understanding.

CHAPTER 1
PRAYING TO OUR DIVINE CREATOR ABOUT *EVERYTHING*

"Pray without ceasing."

—1 Thessalonians 5:17

Life can be discouraging while living in this world that is so full of suffering, tragedy, and disappointment. Even for those of us who are Christ followers, it can sometimes feel like a struggle just to live our everyday lives.

However, there is an immutable truth that helps us rise above all the pain and misfortune: We can pray about *everything*—even about situations that seem impossible to us—because we have a Divine Creator.

God is the all-powerful Creator and our loving Heavenly Father. He created us in His image, and He wants to fellowship with us. Prayer is one of the best ways we can communicate, fellowship, and grow with God. And He wants us to pray without ceasing. However, it can be difficult to pray to Him when we aren't sure who He is and when we don't have a clear idea of just how powerful and loving He is.

The more we explore His amazing, complex, and unfathomable creations, both great and small, the easier it is to pray to Him. After all, He has created the ability for us to hear, see, feel, and think—all processes that require many complex physiological and biological processes—so He certainly has the power to hear our prayers and ease our suffering.

He loves us so much that He sent His only Son to die on the cross for our sins. When we acknowledge His power and His love for us, it enriches our prayer life, which in turn brings us closer to Him and empowers us to live the life He planned for us before we were even born.

As Christians, we pray to God about our own needs, such as for wisdom and guidance. He also wants us to pray for others, and when we do so, we become *intercessors*—people who pray on behalf of others, pleading with God for their needs. The term also refers to Jesus Christ as the ultimate heavenly Intercessor, who continually advocates for believers.

How Prayer Restored My Faith After a Series of Losses

Prayer is one of the most powerful ways we connect with God.

Most of the time, my prayers to Him are full of praise, gratitude, and worship. But there have been other times, when my life came crashing down around me, that I have expressed frustration, and sometimes even anger, at God. The key to staying close to Him is to pray, regardless of whether things are going well or badly.

Next, I am sharing with you a parade of misfortunes that led to the darkest season of my life. When everything I touched seemed to fail, I was angry with God and began to feel hopeless. I prayed to Him anyway, asking Him to help me get through it all. It took some time, but I finally came to terms with my situation. Many times, I had to remind myself that *all* things—the good ones and the painful ones—work together for the good of His Kingdom:

"And we know that for those who love God all things work together for good, for those who are called according to his purpose."

—Romans 8:28

This Scripture is God's promise to believers who love Him and are living according to His calling—those who have placed their trust in Jesus Christ. It assures Christians that even difficult circumstances serve His greater good in our lives. We have to trust Him fully and understand that He will deliver us from our trials. Enduring the difficult times strengthens us.

I hope that sharing my experiences might help you strengthen your prayer life and your relationship with God, especially if you are experiencing a sense of loss.

1. Beginning—And then Losing—My Marriage and My Ministry

When I was pursuing my Master of Divinity (M.Div) degree at Asbury Theological Seminary in Kentucky, I met the woman who eventually became my wife. She was a volunteer in the youth ministry I was leading. We got married after dating for about a year.

After graduating from the seminary, I took a full-time job as the pastor of a church in Denver, Colorado. Our marriage was great for a couple of years.

About five years into our marriage, my wife said she wasn't happy in our relationship. We agreed that divorce wasn't an option because we wanted our young son to grow up with an intact family; also, the Bible calls for married couples not to divorce:

"To the married I give this charge (not I, but the Lord): the wife should not separate from her husband (but if she does, she should remain unmarried or else be reconciled to her husband), and the husband should not divorce his wife."

—1 Corinthians 7:10–11

We went to marital counseling in an effort to save our marriage. After our second child, our daughter, was born, my wife and I worked together to get through the difficult times. We were building a strong foundation for our family—or so I thought. However, just the opposite was true: our marriage was unraveling.

I was devastated that what I had intended to be a lifelong partnership was falling apart. I felt like I was failing my church as a pastor because with our marriage in peril, we weren't leading our congregation optimally. And how could I continue counseling couples effectively if my own marriage was in shambles?

After our son graduated from high school, he joined the US Marines; our daughter graduated from high school two years later. In our daughter's senior year of high school, my wife told me she wanted a divorce. The plan was for us to be together when we broke the news to both our kids. I kept avoiding it, though, because I knew they would be devastated—and because I kept hoping and praying my wife and I could work everything out.

So one day in June 2014, my wife broke the news to them when I wasn't there. I can't blame her because I kept putting it off.

The marriage had dissolved long before the divorce was final in December 2014.

2. A New Business, My Parents' Deaths, and a Tumor

After the divorce, I resigned voluntarily from my position as a pastor because I believed I shouldn't be a pastor if I was divorced. The church leaders did not ask me to resign.

For twenty years, I had preached to our congregation about the importance of family and strong marriages. I had held marriage workshops at the church and taught the biblical principle that marriage—the primary relationship between a man and a woman—is meant to be for life and is a strong foundation on which to bring up children. I felt that, as the head of my household, I was responsible for the success of the marriage. I was responsible to raise my kids in a Christian home and to partner with my wife in living a Christian life.

The divorce destroyed my credibility as a pastor and also as a husband, a father, and a person of faith. I felt like a hypocrite, being divorced and calling myself a pastor.

It was time for me to move to Plan B in terms of a career.

Back in 2006, years before our marriage ended, my wife and I had bought a medical-supply business; the owner was selling it because he wanted to retire. The idea was that my wife would run the business while I continued to pastor the church. However, after purchasing the business, we found out that because she worked for a rehab center and nursing home, it would be a conflict of interest for her to manage it. So I ended up having to operate the business.

The timing on that actually turned out to be pretty good because after the divorce, I had no place to live. In July 2014, I moved my bed into a small storage room, about the size of a bedroom, at the back of the store.

Around the same time our divorce was final in December 2014, I discovered a lump on the right side of my face. At first, it was tiny—about the size of a grain of sand. But then it grew to be about the size of a pea. I went to see a doctor, and he did a biopsy. It was a tumor, and he said he needed to remove it. When I tried to schedule the surgery, I discovered that I no longer had health insurance. My wife had inadvertently canceled my insurance during the divorce proceedings, and I wasn't aware of it.

I applied for health insurance in January, and it took effect in February. By that time, the tumor was the size of a golf ball.

During that difficult time, both my parents died. I inherited my dad's prized Lionel train set and my mom's roll-top desk. I took those treasured heirlooms to my store.

A friend took me to the hospital when it was time for my surgery. The doctors did a skin graft to cover the hole in the side of my face where the tumor had been. After that, I had radiation for a few months. My kids didn't want me living by myself while I was going through those treatments, so I moved into the basement of a friend's house. I didn't stay there long because I really wanted to move back to my room in the store.

3. A Fire Destroys Everything I Own

One of my hobbies was flying remote-control airplanes. One day in January 2018, at the store that was also my home at the time, I was charging the battery in one of my planes. Suddenly, there was an electrical short, and the battery and plane caught on fire. As the blaze erupted around me, I frantically tried to put it out with blankets. According to the video camera outside the building, I was in there fighting the fire for about six minutes.

Then the electricity went off, so the lights went out. The fire was too big for me to extinguish on my own, and because of the heavy smoke, it became hard for me to breathe. So I exited the building, along with a massive billow of dark smoke. I don't remember when I called the fire department. My car was parked against the building, and some guy who was in the parking lot said, "Hey, you'd better move your car, or it's going to burn up."

It was my dad's car. I inherited it after his passing, and of course I didn't want to lose that, too. So I quickly got into the car and drove it away from the building. I got out of the car and stood there, waiting for the fire trucks to arrive, watching everything I had left go up in flames. I didn't realize it, but a local news station was filming the fire and broadcasting it on the local news. Someone from the church I was attending recognized that it was my new business and drove there to see if I was OK. I appreciated that.

I called my kids and told them about the fire. Then the paramedics arrived and made me get in the ambulance because I was covered with soot. They were worried about smoke inhalation because there were a lot of fumes. I stayed in the ambulance—hiding, basically—because I didn't want to face the landlord or anyone else.

The firemen were tossing items from the store out the window so they wouldn't provide further fuel for the fire. Also, they were looking for hot spots underneath all the items. There's a saying about fires: "What the fire doesn't get, the smoke gets. What the smoke doesn't get, the water gets. And what the water doesn't get, the firemen will destroy."

I went outside and walked around to the back of the building, where my storage-room/bedroom window was. All my personal possessions were lying on the ground in a heap of wet ashes. My mom's roll-top desk was destroyed.

Standing in the middle of that charred little room, covered in black soot and ashes from the floor to the ceiling, I thought, *These are the ashes of my memories and my livelihood. They will blend in well with the ashes of my marriage and of my identity as a pastor.*

One of the firemen who was inside the room found a still-intact $5 bill. As he handed it to me through the window, he said, "Hey, look! You didn't lose everything after all." We both laughed.

After the fire was extinguished, I went back there to see if anything was salvageable. My kids were there; we probably shouldn't have entered the building because who knows what chemicals were in there. I found my dad's train set, which was pretty badly damaged, but I retrieved it and still have it today. I also found my violin and a few more items. I dug them out of the ashes and set them aside, with the intent to return the next day and get them.

One of the few mementoes of my wedding that I had held onto was a large, framed caricature that a friend had drawn for us. It burned in the fire. I also had a really special portrait, a painting, of my kids when they were young. It burned in the fire, too. I set them aside anyway, thinking I might be able to restore them somehow.

When I returned the next day, I was surprised to see that the entire building was sealed off, and bright yellow tape labeled "Fire Line—Do Not Cross" was everywhere. The firemen had discovered asbestos, so they wouldn't let me in the building to retrieve the few treasures I had set aside. That was another punch in the gut because the previous day, I had felt lucky that at least I could salvage a few keepsakes. And now I couldn't even do that.

The landlord and I had to pay a hazmat company to come in and dispose of everything. They sealed whatever remnants of my life were left in bags and hauled it off somewhere.

It was around that time that I first thought about writing a book, and I thought an appropriate title would be *FUBAR Faith,* with FUBAR standing for "Fouled Up Beyond All Repair. (I have changed the first word of the actual title to be family-friendly.)

That's how I felt about my life and my faith. I had to face the crushing realization that I was far from perfect as a pastor, a husband, a father, and a businessman. On top of all that, now I had cancer and almost no earthly possessions left to my name. I felt like a complete failure in every area of my life.

For a couple of months after the fire, I lived in a motel. And then the insurance company paid for me to rent an apartment for a few months, while my store was being rebuilt. Thank God I had some insurance on the business, although it was sadly underinsured.

After the insurance claim was settled and I moved into the newly rebuilt store, I invested the insurance money in my business.

Eventually, I bought a modest home not too far from work and got a roommate, to reduce my expenses.

4. The Pandemic Erodes My Business

Just as I was getting back on my feet with my medical-supply business, the COVID-19 pandemic happened. The downturn in customers just about destroyed my business all over again. I went into debt trying to stay afloat. Years later, I'm still paying off that debt.

Through all my losses, I had gone through a cycle of being angry, trying to restore my trust and faith in God, and trying to remain optimistic about the future. But with each loss I endured, my faith eroded a little more, along with my self-esteem.

There were times when I felt so defeated that suicide almost seemed like a good way to escape the pain, humiliation, and sense of failure that overcame me. But the truth is, I would never take my own life. I would never do that to my children. I have dealt with the aftermath of suicide in the past. I believe that suicide doesn't stop the pain; it just transfers the pain to somebody else—and multiplies it. It becomes an unwanted family legacy and impacts future generations. A lot of people are left feeling guilty after a person dies by suicide.

I had a really hard time letting go of my anger—toward people, my situation, and even God. I felt guilty when I expressed anger toward God. I knew that I had to confess it. I also knew He would forgive me, regardless of how many times it happened. I turned to this Scripture many times:

"If we confess our sins, he is faithful and just to forgive us our sins and to cleanse us from all unrighteousness."

—1 John 1:9

My self-doubt became overwhelming and led to a lot of negative self-talk. Once I allowed my negative thoughts to enter my mind, they mushroomed. It was hard to stop them at that point. But as a Christian, I knew, and still know, that self-doubt and fear are not of the Lord. I read this Scripture often, to remind myself that negative thoughts are never from Him:

"Beloved, do not believe every spirit, but test the spirits to see whether they are from God, for many false prophets have gone out into the world. By this you know the Spirit of God: every spirit that confesses that Jesus Christ has come in the flesh is from God, and every spirit that does not confess Jesus is not from God. This is the spirit of the antichrist, which you heard was coming and now is in the world already."

—1 John 4:1–3

---

5. A Brain Tumor Threatens My Very Existence

In 2024, I began experiencing some strange issues with numbness in my left leg and with my overall balance. I would be walking and would suddenly fall down. On Sunday, November 3rd, at 1:00 in the morning, I fell out of bed and landed on the floor. I stayed there until 6:00 a.m., when my roommate woke up. He helped me back to bed. He expressed concern about me but said he needed to leave to go to a volunteering commitment. I assured him I'd be fine by myself, so he left.

Half an hour after he left, I fell out of bed again, hitting my back pretty hard on a set of metal dumbbells on the way down. It hurt a lot, but I could not find my phone to call anyone. I lay there for a couple of hours. My daughter just happened to call me around 8:00 a.m., and I was able to find my phone and answer it. I told her what happened, and she called my son. They drove to my house together and then drove me to the emergency room at the hospital.

Once I got there, the doctor ordered an MRI of my brain. And guess what they found? A brain tumor.

They scheduled me for surgery, but there was no availability for almost a week. So I was in the hospital for five days, during which time I was a fall risk. As I lay there, waiting for the surgery, I didn't care if I died during the procedure. What worried me was that I might come out of the surgery a different person. After all, the personality is stored in the brain.

On Thursday, November 7th, I had surgery, which involved two procedures and took about 12 hours. The first procedure was to manipulate a catheter up through my groin to my brain to try to limit the bleeding that was anticipated to occur during the second procedure, which was to remove the tumor. Thank God the tumor was sort of sitting on top of my brain; it was not enmeshed within my brain like some tumors are. As a result, the surgeon told me it was relatively easy to remove it.

I named my tumor "Roger."

I was in the hospital for another week after the surgery and had appointments with occupational, physical, and speech therapists to improve my coordination, balance, strength, speech, and cognitive abilities. I had trouble speaking; I couldn't speak the words I wanted to say. When I was talking with someone, I would pause, searching for words. The medical team told me that was normal and that over time, as the swelling in my brain subsided, I wouldn't experience that frustration anymore.

After I was discharged from the hospital, I continued to have regular appointments with all my therapists for several months. The medical team told me I was going to have to reprogram my brain to create new pathways, and revive old pathways, to communicate with my legs and the rest of my body. Experiencing that sensation of communication misfires between my brain and my body reinforced to me just how complex yet resilient we are as human beings.

Gradually, I improved—just as the medical team told me I would. I am simply amazed, not only at how complex we are but also at how the human body can heal from serious illnesses, injuries, and tumor invasions.

As annoying, embarrassing, and inconvenient as those initial falls were for me, it's actually good that I hurt my back when I fell because otherwise, I wouldn't have gone to the hospital.

To show just how amazing the process of healing is, I am including three brain-scan images. The first one was taken when I got to the emergency room and doctors discovered Roger. The second image was taken four days after Roger was removed; you can see a void space where the tumor had been. The third scan shows that the void space disappeared, and my brain mostly resumed its normal shape and size. Just amazing!

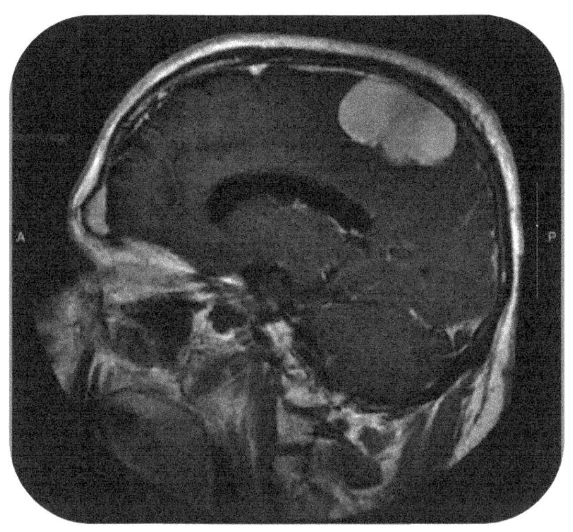

This scan was taken when I arrived at the hospital emergency room on November 3, 2024. The white mass at the top right is my brain tumor, Roger.

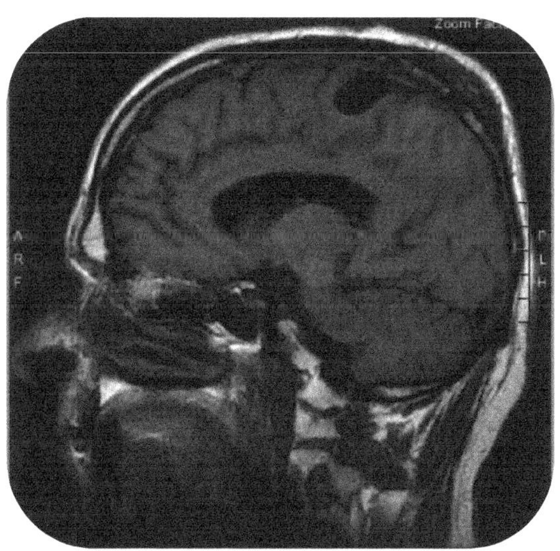

This second scan of my brain was taken right after my surgery in which doctors removed the tumor. The dark spot at the top of the image, to the right of the center, is the void where Roger had been. The gap where the tumor had been was beginning to fill in and heal.

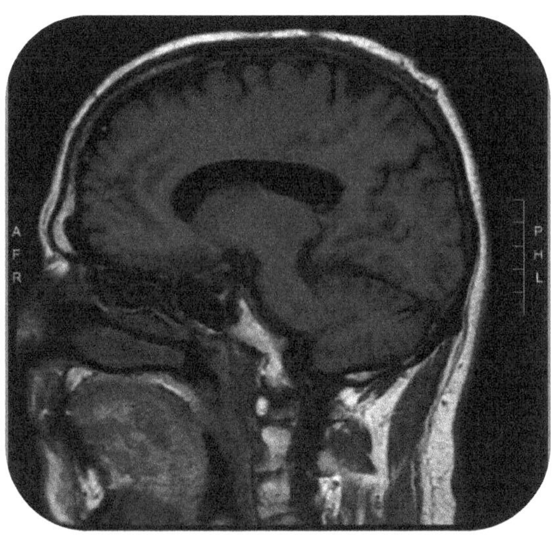

Almost three months after my brain tumor was removed, my brain had resumed its normal shape and size; the black space you saw in the second scan is no longer there.

There is a master design to God's creations, including the ability for us to heal. A surgeon removed my tumor, but God gave my brain the ability to heal—to reprogram the pathways to my body so it can continue to function optimally, despite the temporary intrusion of the tumor. He also gave the surgeon the intelligence and skill to remove my tumor.

God must love me a lot to allow me to survive.

Finding My Way Back to God

Even though I was angry at God, I committed to continue obeying Him in four ways. I would continue to go to worship services at church, serve Him in the church, tithe, and pray. I had always prayed in my mind or out loud, but now, after all these personal losses, I found that I would get distracted and lost when praying. So I began to journal my prayers. I would write my prayers in longhand in Moleskine journals. Over a period of a few years, I filled about seventy of those notebooks with my prayers.

And then I began typing my prayers, and that worked well for me. For some reason, typing my prayers helped me stay focused.

I'd get up early in the morning and go to my computer. I would type the date and the time and start typing my prayers. I have always included what I call the "seven f's" in my prayers: family, friends, fellowship, fiefdom (where you live—city, state, country), future, forgiveness, and faith.

In addition to these practices, I realized that faith was at the core of my journey. Faith kept me moving forward even when I was uncertain or afraid, anchoring me through every challenge and loss. Faith reassured me that there was purpose and hope, even when answers were unclear.

I also journaled about my struggle to continue trusting God and trying to understand why all these things had happened to me. I documented my struggle to restore my faith and trust in Him.

As I continued to pray and journal faithfully, even when I didn't feel like it, my prayer life got better. And I began to reconnect with God. My anger slowly transformed into understanding and renewed trust.

Astronomy Reignited My Sense of Awe About His Power

Around 2023—before I knew I had a brain tumor—I developed an interest in astronomy. I began using a telescope to observe space and studying images from the Hubble and the James Webb telescopes. These images show God's creation on a much grander scale. People are still discovering mysteries of the universe that have never been revealed before.

It is fascinating to me that our solar system has one star, eight planets, five dwarf planets, at least 290 moons, more than 1.3 million asteroids, and about 3,900 comets.[1] What a complex system! Yet all those elements constantly move in relative harmony—just one of the miracles orchestrated by our Master Designer.

When God said, "Let there be light," He created not just the sun, the moon. and the stars; He created the four fundamental properties of physics (gravity, electromagnetism, weak nuclear force, and strong nuclear force) and all the minute building blocks of life and space.

The more I have grown in my Christian faith, and the more I continue to learn about the marvels of science, the more I understand how science and faith complement each other.

First, I believe God created the universe and everything in it—because that's what God's Word, the Bible, tells us—and I believe the Bible:

"Then God said, 'Let us make man in Our image, according to Our likeness; let them have dominion over the fish of the sea, over the birds of the air, and over the cattle, over all the earth and over every creeping thing that creeps on the earth.' So God created man in His *own* image; in the image of God He created him; male and female He created them. Then God blessed them, and God said to them, 'Be fruitful and multiply; fill the earth and subdue it; have dominion over the fish of the sea, over the birds of the air, and over every living thing that moves on the earth.'"

—Genesis 1:26–28, KJV

Second, I believe humankind developed science to enable us to understand how God's creation works and how to live our lives within His universe in a way that glorifies Him. So, with regard to one of the most common debates about science and faith—the

1 "Solar System Exploration," NASA website, https://science.nasa.gov/solar-system/.

evolution-vs.-creation debate—I believe it's not an "either/or" question. God created all living things and gave them the ability to adapt over time.

Seeing star nebulas, galaxies, planets, and stars renewed my awe of God's power, majesty, and master design. I had always known, but had lost sight of, the fact that if God can design and create everything in the universe—from organisms that are so small we can't see them to the entire solar system—then He could certainly help me overcome my problems.

The situations I had been agonizing over suddenly seemed minute compared to all the wondrous miracles God had created. And they were nothing compared to the glory I will someday experience with Him in heaven:

"For I consider that the sufferings of this present time are not worth comparing with the glory that is to be revealed to us."

—Romans 8:18

"Blessed is the man who remains steadfast under trial, for when he has stood the test he will receive the crown of life, which God has promised to those who love him. Let no one say when he is tempted, 'I am being tempted by God,' for God cannot be tempted with evil, and he himself tempts no one."

—James 1:12–13

God allows tragedy in our lives because He wants to build our character, grow our faith in Him, and bring us closer to Him. He wanted to use the tragedies in my life to strengthen my resilience and to remind me to always trust Him. Navigating each tragedy made me stronger as I faced the next one.

> "We rejoice in our sufferings, knowing that suffering produces endurance, and endurance produces character, and character produces hope, and hope does not put us to shame, because God's love has been poured into our hearts through the Holy Spirit who has been given to us."
>
> **—Romans 5:3–5**

The Bible also tells us that all things work together for His purpose:

> "And we know that for those who love God all things work together for good, for those who are called according to his purpose."
>
> **—Romans 8:28**

Eventually, God will use even negative situations in our lives to produce a good outcome for those who follow Him.

My Losses Are All Gains in Christ

A few days after my brain surgery, propped up in my hospital bed, I looked in the mirror at the bandages on my head. I told myself, *Hey, this isn't so bad.*

For a while, I was bedridden and couldn't stand up unless someone put a gait strap around me and helped me up. But I knew I would get stronger, with time. What amazed me was that I had just gone through brain surgery, but there I was, sitting up, talking, and thinking. My biggest worry going into the surgery wasn't that I might die; I was worried about coming out of it with a different personality or with brain damage. But I didn't have either! I was pretty much intact. I could spend more time with my kids! I could reset my life. I could start over, with a new appreciation for all of God's creations.

I realized that I have never starved or been paralyzed. I've never fallen off a roof and broken my neck; I know people who have been through that. I've never had a child born with spinal bifida or cerebral palsy. I know people who have dealt with a lot worse situations than me.

One reason I wrote this book is to help people see that God is still in control, even when everything seems to be collapsing around us. He has a Master Plan, and we simply need to trust and follow Him—regardless of what happens in our lives.

What matters most is that God, our Creator, sent His only Son, Jesus Christ, to die on the cross for our sins. When we accept Jesus as our Personal Savior, we will enjoy eternal life with Him in heaven. That is an incredible miracle and a blessing. When we realize what a gift salvation is, tangible possessions of this world don't matter—at all!

"But whatever gain I had, I counted as loss for the sake of Christ. Indeed, I count everything as loss because of the surpassing worth of knowing Christ Jesus my Lord. For his sake I have suffered the loss of all things and count them as rubbish, in order that I may gain Christ and be found in him, not having a righteousness of my own that comes from the law, but that which comes through faith in Christ, the righteousness from God that depends on faith."

—Philippians 3:7–9

All these experiences in my life—most recently healing from a brain tumor that could have proved fatal—have strengthened my prayer life. I now communicate with God on a deeper level than ever before, and I appreciate, more than ever, His power and His love for me.

I hope that my story will help you overcome any trying situations and/or lapses in faith you are going through. Even if your faith has wavered, please continue to pray, and continue to trust Him. Stay in communication with Him, and in time, your cosmic connection with Him will be restored.

Why I Wrote This Book

I wrote *Cosmic Conversations with Our Creator* because I want to share the awe and reverence I have for God so others can appreciate His magnificence and strengthen their prayer life. I believe a rich prayer life is the key to developing a deeply personal relationship with God.

My hope is that describing how magnificent God is will help you understand the all-powerful and loving God to Whom you are praying so you can pray to Him with more confidence—about *everything*.

The human body is far more detailed than any painting any person could ever paint. And the universe is far more intricate and detailed than any telescope could even capture. Everything we see, and can't see, was created by God. And He is the magnificent One we pray to.

I hope this book encourages you to reach out to God in prayer with a heart that says, "Wow, God is so much more than I've ever thought of before! He created everything, including me, and He wants me to communicate with Him regularly by praying." I have included a sample prayer at the end of each chapter just to give you an idea of how you might pray to Him. I encourage you to modify the prayers as needed so they are relevant to your own personal situation.

Also, I hope that, by demonstrating how science and faith complement each other, those who read this book will experience a renewal of faith in God. When we realize the incredible power it took for Him to create organisms so small we cannot see with the naked eye and galaxies so huge we can't even fathom, I think it is much easier to realize that He can, and does, answer our prayers.

Since becoming a Christian, I haven't always lived a righteous life. And there have been times when I stopped praying because I was angry at God. But He didn't turn His back on me; in fact, He enabled me to recover from having a brain tumor.

I am sharing my personal story in the hope that it might encourage someone else to continue trusting in Him, even during the most difficult moments of life.

About the Book's Title

We live in an unbelievably complex world in which many intricate systems work in concert with one another to sustain life. Some call it a "cosmos."

One definition of *cosmos*, from Merriam-Webster, is "an orderly harmonious systematic universe." From "cosmos," the word *cosmic* simply means "relating to the cosmos." A more specific definition of "cosmic," also from Merriam-Webster, is one I like much better: "Characterized by greatness, especially in extent, intensity, or comprehensiveness."

When we pray, I believe our prayers need to be more than just vague requests to God for this or that, cloaked in doubt and uncertainty. Also, I believe our prayers should be fervent, sincere pleas from the depth of our hearts—not rote prayers we've memorized. Our prayers should be *cosmic conversations with our Creator*! And we need to view prayer as an honor and a privilege, not as a chore.

Our prayers need to express both acknowledgment of, and gratitude for, the countless marvels God has created, from the smallest to the largest. In other words, just like God's creations, our prayers to Him need to be "characterized by greatness, especially in extent, intensity, or comprehensiveness."

The more we understand, acknowledge, and appreciate His greatness and His love for us, the more effective our prayers will be. We will pray about everything because we know and understand that we have a Divine Creator.

The images in this book can never truly convey the brilliance of His creations, but they are a start. I hope they will help you understand and appreciate His magnificent brilliance. I hope they inspire your prayers. And I hope the Scriptures provided in this book reveal to you just how much He loves each of us.

I pray that you will saturate your mind with these passages of Scripture and stunning photos so you can further understand His love for us and why He created us in His own image.

Combining Our Prayers with Actions

Prayer is vital to maintaining a close personal relationship with our Creator. However, He wants us to combine our prayers with actions—to *do* something helpful:

"What good is it, my brothers, if someone says he has faith but does not have works? Can that faith save him? If a brother or sister is poorly clothed and lacking in daily food, and one of you says to them, 'Go in peace, be warmed and filled,' without giving them the things needed for the body, what good is that? So also faith by itself, if it does not have works, is dead."

—James 2:14–17

> "Be doers of the word, and not hearers only..."
>
> **—James 1:22**

Combined with actions, our prayers become catalysts that strengthen our communications with Him and lead us to serve His Kingdom more dutifully.

Often, when a tragedy occurs, such as a school shooting, well-meaning people say, "I will pray for you" or "My thoughts and prayers are with you." Offering such encouragement is kind and thoughtful, but in my opinion, we can do so much more. I believe we are truly serving as ambassadors of our Lord Jesus Christ when we make the effort to take specific actions to help people who are facing hardships.

Instead of simply praying for someone and saying, "Let me know if there's anything I can do," instead, offer to help in a way that you know the people involved would appreciate. Or you could ask them or their loved ones what type of support they need.

My hope is that these prayers will serve as templates you can use to customize your own prayers as needed for your situation. Below my prayers and the action items I plan to take after praying, I have provided space for you to write your own prayers, along with the action items you plan to take after praying about each situation.

Acknowledging His Power Leads Us to Rely on Him

I hope that, in revealing more to you about God's power, His awesomeness, and His love for all humanity, this book will lead you into a closer, more intimate relationship with Him. I hope it reminds you that He is not only a God of judgment but also a loving, merciful God who is with us every moment. I hope this book enhances your prayer life and encourages you to pray about *everything*. I also hope that having *cosmic conversations with our Creator* will remind you to cast your burdens on Him instead of trying to carry them yourself.

A Prayer to Our Holy Father, with Gratitude for Healing

Heavenly Holy Father, Creator of all there is, I come to you, thankful for many things. I am grateful that You enable our bodies to heal. I am also grateful for the skill and knowledge of the medical staff at St. Anthony's Hospital for dedicating their lives to the study and practice of medicine and their use of their skills and knowledge to facilitate healing in their patients.

About a year ago, I was in a pretty dreadful state. I am so thankful that You allowed my body to finally communicate to my brain that something was wrong. Thank You that my kids intervened and got me to the hospital.

I did not know the extent to which my brain tumor would alter my life and minimize my ability to function. As I continue to heal, Lord, I pray that You never let me stop being grateful to the surgeons, nurses, support staff, and others who tended to me during that difficult and scary time.

Help me give back and help others, just as those skilled specialists helped me. Show me how You want me to be of service. Perhaps publishing this book is one way I can help others strengthen their prayer lives. Please let me be open to whatever Your will is for my life.

In Jesus's name I pray. Amen.

Actions I Will Take:

1. Volunteer at a local hospital.
2. Share with others the book *Cosmic Conversations with Our Creator*.

Now I encourage you to write your own prayer. Think about what we have discussed in this chapter, and pour out your heart to Him. What do you want to say to Him? What do you want Him to communicate to you?

____

____

____

____

____

____

____

Actions You Will Take:

1. ________________________________
2. ________________________________
3. ________________________________

CHAPTER 2
RECONCILING SCIENCE AND FAITH

Since the dawn of Darwin's theory of evolution, people have debated whether it is possible for science and religion/faith to coexist. I'm sure this debate will continue until the end of time!

I have been fascinated by these two aspects of our human existence for most of my life. I believe science and faith actually *complement* each other rather than just coexist.

Here's how my fascination with this topic began.

When I started studying biology at the college level, I noticed a tension—a conflict—between the concepts of creation and evolution. People seemed to believe in either one or the other idea of how we came to exist.

Human physiology was one of my favorite courses. My final exam in that course consisted of a single essay question: "Describe the human central nervous system (CNS)." In the process of explaining the highly intricate and complex way the human CNS coordinates communication between our brains and our bodies, I remember sitting at my desk in that quiet classroom, awed at how God created us with such complexity and precision—yet as we go about our daily lives, we're typically not even aware of all that's happening within our brains and bodies and beyond.

In my zoology class, the professor's syllabus began with evolution. On the first day of class, he said, "Here's the deal. The only acceptable answers you give in this class and on tests will deal with evolution. We don't discuss creation here. If you want to pass this class, you have to answer these questions based on the facts of evolution."

I was really discouraged about that at first, but I needed the class to graduate, so I decided to respond to the questions with answers the professor wanted to hear. I don't think I'm alone in this. I think many Christians, when they begin to study evolution, experience the same dilemma. As I learned more about the theory of evolution, I realized there is some truth to it; however, it doesn't happen by itself. Evolution simply *expands* the creative power of God; it doesn't *negate* it.

That experience actually strengthened my faith in a way I wasn't expecting.

While I was going to college, I worked at a hospital for several years. I also served as a volunteer youth director at a church, teaching Bible lessons to kids and accompanying them on trips. During that time, I constantly felt God pulling on my heart to go into the ministry, but I decided to go into the Peace Corps instead. I wanted to use my science degree to help people. While I was in Costa Rica, God kept calling me. I really felt a need to become a pastor. I talked to a missionary who was serving in Costa Rica. He told me, "It sounds like you need to study theology and become a pastor because you are not going to be happy doing science by itself."

While in the Peace Corps, I applied to the seminary the missionary recommended, Asbury Theological Seminary in

Wilmore, Kentucky, about sixteen miles southwest of Lexington. I was accepted there, and three months after I got out of the Peace Corps, I began my seminary studies. It took me more than three years to complete my Master of Divinity (M.Div.) degree, although I prefer the term "biblical theology."

The Mysterious Origin of Time

People have asked me, "How could God have created the heavens and the Earth when Jesus was here just two thousand years ago, yet the Earth has been around for billions of years?"

My response is that time is not a creation of God; it is an invention of man to measure how God's creation works. Whether we're talking about the 24-hour day, the length of the seasons, or the time it takes the planets to orbit the sun, I believe man invented the construct of time to help make sense of it all.

In 2 Peter 3:8, Peter wrote, "But do not overlook this one fact, beloved, that with the Lord one day is as a thousand years, and a thousand years as one day." In other words, time to God is not the same as time is to us. God is outside of time—the past, present, and future are one to Him. While to us, time is temporal, to God, time is eternal.

Sometimes, people try to disprove creation by pointing out imperfections in science. However, God's creations are perfect; what's imperfect is humankind's ability to *understand* God's creations. We have changed scientific theories for centuries because new evidence comes forth, thanks to technology that God has given us the ability to invent. Our understanding of time, like our understanding of the cosmos, is incomplete and imperfect—yet all of God's creations are perfect.

For example, in May 2025, observations by the Hubble Space Telescope confirmed that it takes Uranus 17 hours, 14 minutes, and 52 seconds to complete a full rotation. That is 28 seconds longer than estimates made by NASA's *Voyager 2* spacecraft in the 1980s. Uranus is the seventh planet from the sun; it takes about 84 Earth years for Uranus to orbit the sun. (An *Earth year* is the amount of time it takes for the Earth to complete one full orbit around the sun.)[2]

Every four years, an extra day is added to the month of February to create a "leap year." The purpose of leap years is to keep our Gregorian calendar aligned with Earth's actual orbit around the sun. Earth takes approximately 365.2422 days to complete one orbit, which is slightly more than the 365 days in a standard year. Also, each day does not contain exactly 24 hours, and not all months contain the same number of days.[3]

Did God create the world in seven days, or did it take seven billion years? We can't be sure because, again, time is not God's creation. I believe humankind developed time as a way of measuring God's creations. God is infinite. He is omnipresent, which means that for God, there is no past, present, or future. To me, it doesn't matter how long it took God to create the world and everything in it.

Saint Augustine (354–430 AD) was a North African bishop and one of the most influential Christian thinkers and writers in Western history. He, too, wondered about the construct of time and wrote, "What then is time? If no one asks me, I know what it is. If I wish to explain it to him who asks, I do not know."[4]

---

2 "A Day at Uranus Just Got 28 Seconds Longer," Marcia Dunn, Associated Press, April 7, 2025, https://apnews.com/article/uranus-nasa-hubble-19a4d69e0778704a4fd01e6a6f6bd814.

3 "Doing the Math on Why We Have Leap Day," Lyle Tavernier, NASA, January 16, 2024, https://www.jpl.nasa.gov/edu/news/doing-the-math-on-why-we-have-leap-day/.

4 Dean Buonomano, *Your Brain Is a Time Machine: The Neuroscience and Physics of Time* (W. W. Norton, 2018), 4. Dean Vincent Buonomano is an American neuroscientist, author, and professor of neurobiology and psychology at UCLA. His research focuses on neurocomputation and how the brain tells time. He draws on evolutionary biology, physics, and philosophy to present his influential theory of how we tell, and perceive, time. The human brain, he argues, is a complex system that not only tells time but creates it; it constructs our sense of

Another scientist who has written extensively about time is Dan Falk, a Toronto-based science journalist. In 2018, Buonomano published the book *Your Brain Is a Time Machine: The Neuroscience and Physics of Time*, and Falk interviewed him about the book.

The professor told Falk, "We know that many…forms of timing rely on what's called a 'neural population clock'—that's a circuit of neurons where one neuron can contact and excite another neuron, and another, and another. You can imagine them as falling dominoes. If you have a long line of dominoes, you could use that as a clock, because you could mark time based on which domino is currently falling. So that's how the brain tells time on the scale of milliseconds, using what we call 'neural dynamics.' Neurons make up a dynamical system, and they create spatio-temporal patterns of activity, and we use those spatio-temporal patterns of activity to tell time."[5]

The professor also explained to Falk, "Time sits at the center of a perfect storm of unsolved scientific mysteries involving free will, consciousness, and the unification of relativity and quantum mechanics."[6]

People's beliefs about the concept of time span a wide spectrum, often depending on their beliefs about how we got here.

Einstein's and Darwin's Views of Science and Religion

In 1953, Albert Einstein, one of the most brilliant scientists of our time, published an essay titled "Science and Religion," which

---

chronological flow and enables "mental time travel"—simulations of future and past events.

5 "Does Anybody Really Know What Time Is?" Dan Falk, Nautilus website, July 8, 2025, https://nautil.us/does-anybody-really-know-what-time-is-1223272/.

6 Ibid.

contained this quote: “Science without religion is lame; religion without science is blind.”[7]

Many theologians and scientists have discussed what they believe Einstein really meant, but the general consensus is that science helps us understand the physical structure of the universe, while religion deals with human values, morals, and meanings. We need both science and religion to gain a complete understanding of the universe.

People’s thoughts about the creation-vs.-evolution debate span a wide continuum. At one extreme are *atheists*, who don’t believe in God, and *agnostics*, who don’t believe it’s possible to know if God exists. People in those camps tend to rely only on facts, such as scientific truths—what we can see and measure. Some people believe only in evolution, and others believe only in creation. In between are those who believe that God created all things, as the Bible states, and that over time, living things have adapted.

Scientist Charles Darwin is credited with developing the theory of evolution. He proposed that species change over time through a process called “natural selection,” whereby organisms that have advantageous traits are more likely to survive and reproduce and to pass those traits on to future generations. Many people think Darwin was an atheist or an agnostic, but others say he was more of a “deist,” meaning he believed in a Creator, but not in the traditional sense of God. However, his beliefs weren’t as anti-religious as many believe, and his wife, Emma Wedgewood, was a devout believer.

On January 29, 1839, in a small chapel in the English village of Maer, Emma Wedgwood, a religious 30-year-old woman, married Charles Darwin. She believed firmly in a heaven and a hell, and she believed you had to accept God to go to heaven. He was making

---

7 “Einstein’s Famous Quote About Science and Religion Didn’t Mean What You Were Taught,” Jerry A. Coyne, *The New Republic*, December 4, 2013, https://newrepublic.com/article/115821/einsteins-famous-quote-science-religion-didnt-mean-taught.

notes about a new idea—his theory of evolution—in leather-bound notebooks marked "private." He hesitated to share his thoughts with Emma, but he wanted to be honest with her. He told her of his doubts about the veracity of the Bible and of his growing skepticism about religion. She valued his candor and viewed him as a good and moral man; however, she was sad that they did not agree on such a fundamental topic.[8]

Charles and Emma had ten children together. Three of the children died; the death of their beloved ten-year-old daughter, Annie, especially broke their hearts. That loss strengthened Emma's faith but all but closed the door on God for Charles. They were married for forty-three years. Although they never saw eye-to-eye on the question of the existence of God, they accepted each other's views.[9]

This is an example of how, so often, we allow deep hurt rather than God's truth to shape our view of Him. The deep hurt Darwin felt about losing his children most likely changed his feelings toward God. I think this happens to a lot of people. They lose a loved one, become angry at God, and turn away from Him. We let that hurt define our concept of God. I understand how much they hurt; I have experienced those devastating losses myself. It's normal, and OK, to ask God, "Why did this happen?"

We don't always know why bad things happen, but again we do know that everything works together for His will:

---

8 "The Darwins' Marriage of Science and Religion," Deborah Heiligman, *Los Angeles Times*, January 29, 2009, https://www.latimes.com/la-oe-heiligman29-2009jan29-story.html.

9 Ibid.

> "And we know that for those who love God all things work together for good, for those who are called according to his purpose."
>
> **—Romans 8:28**

I believe that when someone is suffering a loss, it's not the time to beat them over the head with theology. Instead, it is an ideal time for prayer and compassion.

Trying to Make Sense of Darwin's Theory of Evolution

Many people have attempted to prove or disprove Darwin's theory of evolution. One of them is Michael Behe, PhD.[10]

In 1996, Behe, a biochemist, published the book *Darwin's Black Box: The Biochemical Challenge to Evolution*. In it, Behe launched what people have called the "intelligent design movement"—the argument that nature shows evidence of design, beyond Darwinian randomness. Ten years later, in 2006, Behe released the second edition of the book. In the new afterword for that update, Behe explains that the complexity that microbiologists discovered had dramatically increased since he first published the book. He said that complexity continued to challenge Darwinism, and evolutionists have had no success in explaining it.[11]

---

10 Michael Behe, PhD, is a Professor of Biological Science at Lehigh University in Pennsylvania, where he has worked since 1985. From 1978 to 1982, he did postdoctoral work on DNA structure at the National Institutes of Health. He earned his PhD in biochemistry from the University of Pennsylvania. His current research involves delineation of design and natural selection in protein structures. In addition to teaching and research, he works as a senior fellow with the Discovery Institute's Center for Science & Culture.

11 *Darwin's Black Box: The Biochemical Challenge to Evolution*, Michael J. Behe, Free Press, 2006, second edition.

His book provides a compelling argument for the fact that evolution is simply one component of creation—not the sole origin of life.

In writing about genetic mutations, Behe states one of the key components of evolution is that *natural* mutations create stronger organisms with a better chance for survival. However, when humankind has created and implemented mutations, they have, for the most part, resulted in weaker organisms. For example, researchers were able to genetically manipulate the fruit fly to have bigger wings, but then the flies' muscles weren't big enough to flap the wings. So the scientists went back and genetically engineered a fruit fly to have stronger muscles that could flap wings. However, the wings were too small, and the movement shredded the wings.

Behe argues that the existence of multiple complex systems rules out evolution because they are "irreducibly complex"—they cannot function if they are missing just one of their many parts.

He says natural selection works on small mutations just one component at a time. If dozens or even hundreds of distinct proteins, precisely fashioned, are required to make a functional cilium, how could natural selection slowly and patiently craft them, one at a time, while waiting for the complex function of ciliary movement to emerge? It couldn't. So, according to Behe, the hypothesis that the cilium was produced by evolution is therefore disproved. If evolution did not make the cilium, then "intelligent design" must have. In his book, Behe writes, "Life on Earth at its most fundamental level, in its most critical components, is the product of intelligent activity."[12]

---

12 "Review of Michael Behe's *Darwin's Black Box*," Kenneth R. Miller, National Center for Science Education, *Creation/Evolution Journal 16*, no. 2 (September 19, 2008), https://ncse.ngo/review-michael-behes-darwins-black-box.

I often wonder how Darwin's theories might have been different if he had been able to use an electron microscope or a Hubble telescope. Also, I wonder if his theories about evolution and creation would have been different if he had not suffered the devastating losses of several children.

The diversity in our ecosystem shows God's brilliance and creativity. He created all types of living things—from microscopic organisms to huge mammals—and a complex yet delicate ecosystem that enables them to coexist on the Earth. He also created everything in space and beyond, including planets and moons, star nebulae, asteroids, and galaxies, not to mention gases, molecules, chemical bonds, and physical forces like gravity, electromagnetism, and both the strong and the weak nuclear forces.

Now that we have explored how faith and science complement each other, I want to share how acknowledging His power is a critical key to improving your prayer life—and therefore deepening your personal relationship with Him.

A Prayer to Our Holy Father, Acknowledging His Power

Holy Father, Creator of the heavens and the Earth, sustainer of all that is there—not only of the stars, the Earth, the moon, and the sun, but of gravity and the chemical bonds that hold planets together and the nuclear fission that gives stars their light, I come to you in prayer with thanksgiving and awe. King of Kings, Lord of Lords, I worship You and thank You for Your power, greatness, and sovereignty. The universe moves the way You commanded it. What am I, O Lord, that You are mindful of me? I'm not even a dot in the vast infinity of space on a tiny blue marble in an ocean of darkness. Yet here I am, living in the light of the cross, with the hope of Your resurrection, because You love me.

You sent your Son to die on the cross for my sins. You gave me free will and sent your Son to restore me to a right relationship with you. What great love, what great power! What awesomeness there is in You that You still think of me.

Now, Heavenly Father, I come to You to pray for my family, my friends, and fellowship with other Christians. Because You created the vast expanse of space and every living thing, You know the hearts of the people I pray for.

In Jesus's name I pray. Amen.

Action I Will Take:

1. Read more scientific books and articles, as well as Scriptures, to better understand God, our Creator.

Now I encourage you to write your own prayer. Think about what we have discussed in this chapter, and pour out your heart to Him. What do you want to say to Him? What do you want Him to communicate to you?

Actions You Will Take:

1. ______________________
2. ______________________
3. ______________________

CHAPTER 3
ACKNOWLEDGING HIS POWER, WHICH WILL FUEL OUR PRAYERS

Solar prominence, solar flare, and magnetic storms. Influence of the sun's surface on the Earth's magnetosphere. Elements of this image furnished by NASA.

According to NASA, solar flares are the most powerful explosions in our solar system. The largest solar flare can have as much energy as one billion hydrogen bombs! The energy produced by a solar flare can reach the Earth in about 8 minutes. Although we are protected from solar flares here on Earth, they can disrupt communications in the upper atmosphere. They also can affect satellites and spacecraft outside our planet's protective shield.[13]

When I look at the image above of a solar flare and think about these facts NASA has presented, I am once again amazed by God's creative power and wisdom. Not only did He create the

---

13 "Solar Storms and Flares," NASA website, https://science.nasa.gov/sun/solar-storms-and-flares/.

universe and everything in it; He also created the Earth's magnetic field and atmosphere as a protective barrier against the damage that such a powerful explosion could cause.

This is the power of the One we pray to; we must not lose sight of His creativity, wisdom, or power. This is the same God whose grace and love for us prompted Him to send his only Son, Jesus Christ, to die for our sins. Human words cannot begin to describe the magnitude of His power and love.

Both as a Christ follower and as a pastor, I have witnessed firsthand the struggles some people have when trying to pray to God. Often, people aren't sure Whom they are praying to. They might perceive God our Father, our Creator, as a stern taskmaster, or they might not have any idea what God is like. As a result, when they pray, they aren't sure He is really there, much less that He will hear them or answer their prayers. And when they have doubt, they aren't sure He actually has the power to change things. That typically results in a reluctance to pray or half-hearted prayers.

The more clearly we can *see* how powerful God is and *feel* how much He loves us, the more effectively we can pray. When we acknowledge just how powerful God is, and how much He loves us, it makes it easier to pray to Him all the time and about *everything*. It helps us believe He will answer our prayers, according to His will.

Pray Without Ceasing, and Believe That He Hears You

We pray to ask Him for forgiveness when we sin, to worship and thank Him, to seek His will for our lives, and to grow in our faith. The Bible encourages us to pray regularly, without ceasing, and in all situations:

"Rejoice always, pray without ceasing, give thanks in all circumstances; for this is the will of God in Christ Jesus for you."

—1 Thessalonians 5:16–18, KJV

"Do not be anxious about anything, but in everything by prayer and supplication with thanksgiving let your requests be made known to God. And the peace of God, which surpasses all understanding, will guard your hearts and your minds in Christ Jesus."

—Philippians 4:6–7, KJV

When we pray, we must believe He will answer our prayers. Doing so glorifies Him and strengthens our faith in Him:

"And this is the confidence that we have toward him, that if we ask anything according to his will he hears us."

—1 John 5:14, KJV

"Therefore I tell you, whatever you ask in prayer, believe that you have received it, and it will be yours."

—Mark 11:24, KJV

Now, God does not always answer our prayers the way we hope He will, and often not immediately. That doesn't mean God is ignoring us or doesn't hear our prayers. It simply means He is still working on the issue or has another plan in mind for us or for the rest of His creation. Often, what we are praying for may not align with His will. We must always pray for His will to be done:

"And this is the confidence that we have toward him, that if we ask anything according to his will he hears us."

—1 John 5:14, KJV

When I was a pastor, people used to ask me why God allows bad things to happen to people, including His followers. Along those lines, you might ask, "If God created all things, then didn't He give you the brain tumor?"

My answer is that He allowed the tumor to appear and grow in my brain. I don't know why, and I likely will never know why. It has occurred to me that one reason I pushed forward with this book was because I had the brain tumor. Maybe He allowed me to have the brain tumor as a way to encourage me to write this book.

"The Fall" of Adam and Eve Is the Source of Human Hardship

Many of the calamities that happen as part of our human existence are a result of "the fall"—Adam and Eve's disobedience toward God and following Satan's urging to eat the forbidden fruit in the Garden of Eden. In Genesis 2:19, we read God's command to Adam. He said, "And the Lord God commanded the man, 'You are free to eat from any tree in the garden; but you must not eat from the tree of the knowledge of good and evil, for when you eat from it you will certainly die.'"

But then the evil one convinced Eve it would be OK to eat the fruit from that tree. She and Adam believed the lies of the evil one over God's truth. God designed humankind with *free will* so we could make our own decisions. Once Adam and Eve disobeyed God, humankind was destined to experience hardships. This demonstrates the powerful negative outcomes that can result when we take our focus off God and doubt His sovereignty. The more we are in communication with God, the less likely we will be to believe Satan's lies.

Scripture in the book of Genesis reveals what God told Adam in the Garden of Eden:

"The Lord God took the man and put him in the garden of Eden to work it and keep it. And the Lord God commanded the man, saying, 'You may surely eat of every tree of the garden, but of the tree of the knowledge of good and evil you shall not eat, for in the day that you eat of it you shall surely die."

—Genesis 1:15–17

However, God didn't tell Adam *why* he and Eve should not eat that specific fruit.

In our lives as well, we may not know *why* God asks us to do or not do something—but we still need to obey Him. He has a master plan for each one of our lives, and we need to trust Him. Satan tells us to do what feels good, and he makes the temptation to stray from God's will incredibly appealing. This is why it is imperative that we stay in close relationship with God at all times—with prayer being an effective way to do that.

God's Power Enables Us to Resist the Evil One

The subject of Satan and evil is sometimes controversial. Although some people don't want to believe Satan is constantly trying to cause us to sin, the Bible tell us this is true. Just as Satan tricked Adam and Eve in the Garden of Eden, he is still alive and well today, seeking to separate us from our Lord:

"Be sober-minded; be watchful. Your adversary the devil prowls around like a roaring lion, seeking someone to devour."

—1 Peter 5:8

God wants us to pray for spiritual strength when Satan—the evil one, the fallen angel—tempts us to stray from the path of righteousness. God is the only One powerful enough to help us resist Satan's evil schemes. Recognizing how powerful He is gives us assurance that He will protect us from evil, and praying to Him helps us "put on the whole armor of God":

"Finally, be strong in the Lord and in the strength of his might. Put on the whole armor of God, that you may be able to stand against the schemes of the devil. For we do not wrestle against flesh and blood, but against the rulers, against the authorities, against the cosmic powers over this present darkness, against the spiritual forces of evil in the heavenly places."

—Ephesians 6:10–12

Because this is a controversial subject, I have not shared this next event about my personal encounter with a demon with many people, although I did share it with my kids. I hesitated to include the story in this book because I don't want people to think I'm nuts. On the other hand, it is a compelling testament to the fact that evil exists in this world, and Jesus is the only One powerful enough to defeat Satan and his influence.

When I was attending college at New Mexico State University in Las Cruces, New Mexico, I sometimes went to St. James Episcopal Church. Founded in 1875, it is the area's oldest Episcopal congregation, and it was the first Protestant church in the Mesilla Valley. The church was open all the time, and I often went there to pray, sometimes late at night.

When I was the Youth Director at the church I was attending, there was a girl in the congregation whose dad was having an affair. The situation was devastating her. One night, I decided to spend an entire night at St. James, praying for the family and reading Scripture. I started about 10:00 p.m. I would get down on my knees and pray that God would provide the girl with comfort and strength and that her dad would turn away from his sin and restore his marriage. Then I would walk around among the pews to stay focused, pray out loud, read Scripture out loud, and claim God's power over the situation.

After about four hours, I was pretty tired. So at about 2:00 a.m., I lay down in front of the altar with my head on the carpeted step and my Bible laying on my chest. I closed my eyes and continued to pray amid the silence. All of a sudden, I felt the presence of an unfriendly force and heard something growling and gnarling. I opened my eyes but saw nothing. I knew immediately that this was a demon, an evil force. I was so scared that I could not move. I don't know how long I lay there, frozen in fear. Finally, I was able to whisper "Jesus." The snarling stopped, and the entity disappeared.

I prayed out loud, "God, I am really scared. Please protect me and give me strength."

At that very moment, the front door to the church opened. A man walked in and sat in a pew at the back of the church—at 2:00 in the morning! We nodded to acknowledge each other. He stayed there for about five minutes, praying. We never spoke, but his presence brought me great comfort. I continued to pray until the sun came up.

When we pray, either for ourselves or as an intercessor for others, it is a threat to Satan. We must actively resist him:

"Submit yourselves therefore to God. Resist the devil, and he will flee from you."

—James 4:7

Praying and staying in close communication with our loving Father is our best, and only, defense against evil in our lives. I learned firsthand that the power of Jesus's name causes the devil to flee.

We Are to Praise Him Even During Hardships

Often, when we face hardships in our lives, it is frustrating when we can't understand why. However, even when we don't know *why* things happen, we are to praise Him anyway. I know this can be difficult to do; it's something I have to work on in my life.

Wherever you stand in your relationship with God, I encourage you to pray because it enables you to communicate with Him, and that opens the door for deepening your relationship with Him.

As I mentioned earlier, to guide my prayers on a regular basis, I pray for what I call the "seven F's": family, friends, fellowship, fiefdom (where you live—city, state, country), future, forgiveness, and faith.

God tells us not only to confess our sins to one another but also to pray for one another:

"Therefore, confess your sins to one another and pray for one another, that you may be healed. The prayer of a righteous person has great power as it is working."

—James 5:16

The King James Version of the Bible states the verse like this:

"Confess your faults one to another, and pray one for another, that ye may be healed. The effectual fervent prayer of a righteous man availeth much."

—James 5:16, KJV

God has a plan for each one of our lives, and each one of those plans fits into His Master Plan. Everything works together to glorify His Kingdom, even if we do not understand why or how. Again:

"And we know that all things work together for good to them that love God, to them who are the called according to *his* purpose."

—Romans 8:28, KJV

Even if it doesn't seem like your prayers have been answered, or if they haven't been answered the way you want, please don't give up. Your prayers are never wasted.

I pray for loved ones who have not accepted Christ as their Savior. I have done so for decades. Sometimes it's discouraging that they have not trusted in Jesus yet, but I cannot give up. Who knows—someday, when my loved ones are on their deathbeds, their hearts will stop, but their brains will continue functioning for about seven minutes. (More on that later.) I pray that, during those final minutes of their lives, they will remember all those times I've told them that God loves them and that all they have to do is confess their sins to Him and accept Him as their Savior. Then they can spend eternity in heaven with God—and with me.

When we understand His power, we also recognize His sovereignty to act on our prayers in a way that is consistent with His will and his desire. God created humankind—men and women—in His image, and He wants us to seek His wisdom and stay in fellowship with Him. One way we do that is through prayer.

However, I know it can be really hard to pray sometimes.

When we recognize how powerful He is and how much He loves each one of us, it becomes easier to pray to Him "in all things"—for anything and everything. When we realize that He loves us so much that He sent his only Son to die on the cross for our sins so we could spend eternal life with Him in heaven, then it becomes easier for us to engage in cosmic conversations with our Creator—fervent, heartfelt, confident prayers to seek His forgiveness, guidance, protection, and wisdom and also to praise, worship, and thank Him for everything He has given to us.

Jesus Tells Us How to Pray

You've probably heard of "the Lord's Prayer." It appears in two forms in the New Testament. A shorter version is in Luke 11:2–4, and a longer version, part of Jesus's Sermon on the Mount, is in Matthew 6:9–13:

“Pray then like this:

‘Our Father in heaven,
hallowed be your name.
Your kingdom come,
your will be done,
on earth as it is in heaven.
Give us this day our daily bread,
and forgive us our debts,
as we also have forgiven our debtors.
And lead us not into temptation,
but deliver us from evil.’”

—Matthew 6:9–13

Note that this passage says, “like this.” This means the prayer is an example, a template, that Jesus gave us to teach us how to pray. We don’t need to use these exact words.

Jesus revealed this prayer when a disciple requested that He teach His followers to pray. The prayer begins with acknowledging His greatness and then proceeds with specific requests that will benefit the one who is praying, as well as His Kingdom.

The prayer demonstrates that when we pray to Him, we need to acknowledge His greatness, pray for His will to be done, and ask to be delivered from temptation.

Does God Hear the Prayers of Unrighteous People?

Earlier in this chapter, I included the Bible verse James 5:16, which says, “Therefore, confess your sins to one another and pray for one another, that you may be healed. The prayer of a righteous person has great power as it is working.”

A lot of people read that verse and think, "Of course God would answer a righteous man's prayers. But I haven't always been righteous. Does God hear the prayer of an unrighteous person?"

The answer is yes!

People often cry out to God when they're in horrible situations—often circumstances that are self-made—even if they say they don't believe in Him. And if they ask God to come into their lives and forgive them of their sins—that's how they are redeemed to a right relationship with Christ. He hears our prayers. We are all fallen and have come short of the glory of God—yet He hears our prayers anyway.

The only thing that makes a person righteous is Jesus's death on the cross. A person who accepts Jesus as his or her Savior becomes righteous.

Mark 9:14–27 tells the story of a man who comes to Jesus, asking Him to heal his son. Verses 17 and 18 say the son "has a spirit that makes him mute. And whenever it seizes him, it throws him down, and he foams and grinds his teeth and becomes rigid." In verse 22, the father says to Jesus, "If you can do anything, have compassion on us and help us."

In verse 23, Jesus replies, "If you can! All things are possible for one who believes."

The father of the child says to Jesus in verse 24, "I believe; help my unbelief!"

This is one of the most critical prayers in the Bible. Many people pray these words—me included—when suffering in our lives, and in the world, becomes overwhelming and difficult to understand: "God, I don't understand why this is happening! Help my unbelief."

In this Scripture, the father's belief was wanting. *Our* belief is often wanting. We need to pray for God to correct our unbelief. Jesus helped that father, even though his faith was lacking.

Most people say prayers that are not a cry of faith but rather a cry for help: "God, I am a sinner. Save me. Help me."

But what if someone has lived a life of depraved, gruesome, and unimaginable sin and has not accepted Jesus as his or her personal Savior? Will God hear that person's prayer and allow him or her to be saved? The answer is yes:

"For 'everyone who calls on the name of the Lord will be saved.'"

—Romans 10:13

Back in the 1700s, John Wesley's theology became the basis for the Methodist church. Wesley identified three types of grace. Being aware of these three different types of grace can help us understand why God answers the prayers of people who have not yet accepted Him as their personal Savior.

The three types of grace are *prevenient grace*, which is God's active presence in people's lives before they even sense the divine at work in their lives (before they are believers); *justifying grace*, through which all sins are forgiven by God; and *sanctifying grace*, which allows people to grow in their ability to live like Jesus as they mature in their walks with God.

Sometimes, we try to compare sins, deeming one type of sin "worse than" another. However, God's grace is so powerful and inclusive that He forgives *all* our sins. When He sent His only Son, Jesus Christ, to die on the Cross, that ultimate sacrifice wiped away all sins of mankind—no matter how egregious they might seem.

Notorious serial killer Jeffrey Dahmer killed and dismembered seventeen men and boys between 1978 and 1991. He was arrested on July 22, 1991, and was sentenced to fifteen consecutive life terms. On November 28, 1994, a fellow inmate beat Dahmer to death.[14]

Bob Ross, Jr., was a staff writer for *The Oklahoman* at that time and spoke with Curt Booth, a member of the Crescent Church of Christ in Oklahoma, about the role Booth played in Dahmer's conversion. Ross reported that Booth usually ministered to inmates in prisons close to his home. However, in April 1994, Booth saw Dahmer on TV saying he wished he could "find a little peace." Booth set out to introduce Dahmer to Jesus Christ.[15]

Booth sent Dahmer a Bible correspondence course that taught the steps to salvation. Dahmer mailed the answers back and thanked Booth for the course. And then he told Booth he wanted to be baptized but wasn't sure it was possible because the prison he was in did not have a baptismal tank. Booth contacted Roy Ratcliff, minister of the Madison Church of Christ in Wisconsin. Ratcliff set up weekly Bible lessons with Dahmer and baptized him on May 10, 1994.[16]

When we consider Dahmer's apparent conversion to Christianity, it's easy to shrug it off as impossible or maybe just a rumor. But the truth is, if Dahmer did accept Christ as his Savior—and credible witnesses say he did—then God forgave him of his sins:

"If we confess our sins, he is faithful and just to forgive us our sins and to cleanse us from all unrighteousness."

—1 John 1:9

---

14 "How Police Caught Jeffrey Dahmer," Olivia B. Waxman, *TIME*, July 22, 2016, https://time.com/4412621/jeffrey-dahmer-cannibal-murderer-25th-anniversary-arrest/.
15 "Did 'Jailhouse Religion' Save Jeffrey Dahmer?" Bobby Ross, Jr., *The Christian Chronicle*, August 1, 2010, https://christianchronicle.org/did-jailhouse-religion-save-jeffrey-dahmer/.
16 Ibid.

We are all sinners:

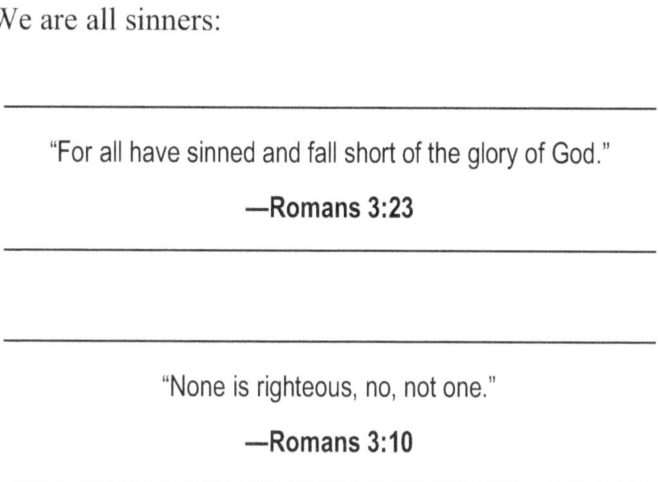
"For all have sinned and fall short of the glory of God."

—Romans 3:23

"None is righteous, no, not one."

—Romans 3:10

This shows us just how powerful God's grace is. If God can hear and respond to the prayers of Jeffrey Dahmer in the midst of his fallenness, wouldn't He also hear and respond to our prayers in the midst of our fallenness?

Sometimes I am unhappy with the way my life turned out, especially after decades of praying for something better. I realize, though, that God doesn't hate me or push me away or remove His grace from me. He loves me, and He listens to my prayers and responds to them, even in my weakest moments—even though I am a sinner.

A Prayer to Our Holy Father, Asking Him to Help Our Unbelief

Heavenly Father, You know there are times in my life when my faith in You is powerful, and I believe everything You say in Your Word. I believe in Your power, grace, and mercy. However, there are times when I struggle, when my faith wavers and weakens. I do not understand how You can allow birth defects, drug addiction, natural disasters, and other calamities. When I question You, please help my unbelief. Please strengthen my belief so it is unwavering.

In Jesus's name I pray. Amen.

Actions I Will Take:

1. Give to, and volunteer to serve with, a Christian organization that supports those in need.
2. Go on a mission trip.
3. Sponsor a child from another country through organizations such as Save the Children and World Vision.

Now I encourage you to write your own prayer. Think about what we have discussed in this chapter, and pour out your heart to Him. What do you want to say to Him? What do you want Him to communicate to you?

____

____

____

____

____

____

____

____

____

Actions You Will Take:

1. ______________________________
2. ______________________________
3. ______________________________

CHAPTER 4
COMMUNICATING TO GROW CLOSER TO HIM

"In these days, he went out to the mountain to pray,
and all night I continued in prayer to God."

—Luke 6:12

A 3D rendering of *Voyager* Space Probe approaching Planet Saturn.

In the summer of 1977, twin spacecraft were launched in Cape Canaveral, Florida—*Voyager 1* and *Voyager 2*. *Voyager 2* was actually launched first, followed by *Voyager 1* two weeks later. Their multifaceted mission for scientific discovery included close fly-bys of Saturn and Jupiter.

Voyager 1 was the first manmade spacecraft to cross the *heliosophere*, which is the boundary where the influences from outside our solar system are stronger than those from the sun. It also was the first human-made object to venture into interstellar space. At Saturn, *Voyager 1* found five new moons, a new ring called the G-ring, a thin ring around Jupiter, and two new Jovian moons called Thebe and Metis.[17]

Voyager 2 is the only spacecraft to study all four of the solar system's giant planets at close range. It was the first human-made object to fly past Uranus and Neptune. This spacecraft discovered a fourteenth moon at Jupiter; ten new moons and two new rings at Uranus; and five moons, four rings, and a "Great Dark Spot" at Neptune.[18]

In addition, both *Voyagers* have a secondary communication mission that involves the "Voyager Golden Record." In essence, it's a time capsule.

Before NASA launched the spacecraft, its engineers placed in both of them a device designed to communicate a story about our world to extraterrestrials! The message is carried by a phonograph record—a 12-inch gold-plated copper disk that contains sounds and images selected to portray the diversity of life and culture on Earth. Instructions for how to play the record are included.

Also on the disk are symbols that show the Earth's relative position to several pulsars, greetings in more than 55 languages, 90 minutes of music across a number of genres, 115 images of life on Earth, and 35 sounds of Earth. They include whale sounds, as well as the sound of human laughter.

Voyager 1 is currently about 15 billion miles away from Earth. Its flight data system collects information from the spacecraft's science instruments and bundles it with engineering data that reflect its operational health. Mission control on Earth receives those data in binary code—a series of ones and zeroes. It is a

---

17 "*Voyager 1*: The Most Distant Human-Made Object," NASA, https://science.nasa.gov/mission/voyager/voyager-1/.
18 "*Voyager 2*: First to Visit All Four Giant Planets," NASA, https://science.nasa.gov/mission/voyager/voyager-2/.

sophisticated, complex communication system that long-ago ancestors might never have dreamed was possible. [19]

In November 2023, the flight data system's telemetry modulation unit malfunctioned. It began sending a repeating code pattern that scientists could not decipher. Without effective communication, that extremely expensive spacecraft was essentially useless. After several months, scientists discovered that a single chip responsible for storing part of the system's memory, including some of the computer's software code, wasn't working properly. They made a clever code modification, and the system's communications were restored.

I see a strong parallel between the way we humans correct abnormalities in technology and the way God heals us. Just as those scientists corrected the tech problem by modifying some code, God enabled my brain to create new pathways after my tumor was removed through a process called *neuroplasticity*, where surviving neurons form new connections to compensate for damaged tissue.

Good Communication Can Win or Lose a War

After the Allied forces' successful D-Day invasion of Normandy, France, on June 6, 1944, it seemed like World War II was over. However, six months later, on December 16th, in the bitter cold of winter, Adolph Hitler made a final attempt to regain control of the Western Front. It was the largest engagement ever fought by the US Army during World War II. The Allies ultimately won the battle, stopping the German offensive and marking the beginning of the end for the Third Reich.

---

19 "*Voyager 1* Is Sending Data Back to Earth for the First Time in 5 Months," Ashley Strickland, ABC 7 Chicago, April 24, 2024, https://abc7chicago.com/post/where-is-voyager-1-now-location-how-far-nasa/14725631/.

Named "The Battle of the Bulge," this conflict was extremely costly for both sides, with more than 81,000 Allied casualties and 100,000 German casualties. The depletion of Germany's armored forces and air power contributed to the country's eventual defeat. Many historians consider the "Battle of the Bulge" the greatest battle in American military history.[20]

Communication problems plagued some of the troops, particularly the 106th Golden Lions Division. The mostly inexperienced soldiers arrived in the Ardennes (in the southern part of Belgium, stretching into Germany and France) on December 11th and were ordered to cover a large section of the US line in a rugged area of Germany known as Schnee Eifel.[21]

Shortly after the German attack began, the commander of the 106th, Major General Alan W. Jones, became worried that the flanks of his 422nd and 423rd regiments were too exposed. He called Lieutenant General Troy Middleton to request that they be withdrawn, but the connection was bad, and they were unable to communicate effectively. Jones came away from the call incorrectly believing that Middleton had ordered him to keep his troops in position.[22]

Unfortunately, German attackers soon encircled the two regiments and cut them off from any support. The GIs were low on ammunition and under heavy artillery fire. About 6,500 of them were forced to capitulate in one of largest mass surrenders of US troops during World War II. After the defeat, General Jones exclaimed, "I've lost a division faster than any other commander in the US Army!"[23]

---

20 "The Battle of the Bulge, 1944–1945," US Army website, https://www.army.mil/botb/.
21 "8 Things You May Not Know About the Battle of the Bulge," Evan Andrews, History.com, last updated February 26, 2025, https://www.history.com/articles/8-things-you-may-not-know-about-the-battle-of-the-bulge.
22 Ibid.
23 Ibid.

This loss occurred, in part, because of poor communication.

Effective Communication Is Critical in Every Aspect of Our Lives

Effective communication is critical for our survival, our social health and relationships, and every other aspect of life—not just in wartime, but also in peacetime.

Try running a business without communicating to your customers, suppliers, and staff. If you did, you wouldn't be in business very long. And try maintaining a healthy friendship or relationship without communicating well. It's not likely to happen.

Corporations spend countless hours and astounding amounts of cash to come up with the perfect slogan to communicate how their products can benefit their buyers. Nike has the three-word slogan "Just do it," which encourages people to be active and stop making excuses. When I was a kid, we all knew the slogan "breakfast of champions" referred to Wheaties cereal, and "Betcha can't eat just one" referred to Lay's potato chips. Perhaps a three-word slogan for this book could be, "Pray about everything."

It's impossible to run a political campaign without communicating effectively with campaign staff, potential voters, fellow politicians, and even your opponent.

For first responders, good communication with their base can be a life-or-death matter when they're trying to find a person who is in potential danger. Each second is critical, and without quick and accurate communication, someone could die. Police officers need good communication to know what situation they are walking into, what crime has been committed, and what the suspect looks like. Without it, again, somebody could die and/or a criminal could escape.

In the past decade, communication technology has grown by leaps and bounds, and now we can communicate with people around the world instantaneously. It's simply mind-boggling!

According to a November 2025 study from Pew Research, about 98 percent of Americans have a cell phone; of those, 91 percent own a smartphone.[24]

It doesn't matter whether someone is in a personal or business relationship or whether someone is a first responder or a slow responder—communication is critical. Without it, valuable time and resources can be wasted. Supplies can be misdelivered or not delivered at all. Relationships can crumble, and battles can be lost.

One study ranked poor communication between spouses as the fifth reason, in a list of nineteen, why couples divorce. The researchers wrote, "A breakdown in the lines of communication is one of the biggest predictors of divorce. Couples who don't communicate well cannot resolve issues together and tend to suffer more misunderstandings and hurt feelings than those who have learned how to resolve conflict respectfully."[25]

Not all communication is verbal. It can be written in a text, email, or letter. It can be physical, such as holding hands, and it can be visual, like a smile. By whatever means, good communication is critical to the survival of a couple, a church, a community, and even a country.

When we pray openly and honestly with God, we are risking a lot. However, God never promised me a perfect life. Yes, life is unfair, and we often struggle, but then God speaks in that still, small voice and lets us know He is with us.

---

24 "Mobile Fact Sheet: Tech Adoption Trends," Pew Research Center, November 20, 2025, https://www.pewresearch.org/internet/fact-sheet/mobile/.
25 "Causes of Divorce: 19 of the Most Common Reasons," Brette Sember, JD, Divorce.com, March 15, 2024, https://divorce.com/blog/causes-of-divorce/.

Prayer Is a Two-Way Dialogue with Him

The title of this book, *Cosmic Communication with Our Creator*, is how I envision the most powerful communication we can have with Him. He wants us to communicate with Him on a very personal level. He wants us to know Him, love Him, and trust Him enough to pray about *everything*. Often, people think prayer is just a one-way request—asking Him for help, for example. But it is so much more! Prayer is a two-way conversation with God, a dialogue.

God longs to communicate with us. Through this divine interaction, we find comfort, guidance, wisdom, purpose, and conviction to do the right thing. He wants us to learn from Him:

> "I will instruct you and teach you in the way you should go; I will counsel you with my eye upon you."
>
> **—Psalm 32:8**

As we pray, we need to listen to Him with receptive and obedient hearts. We need to ask the Holy Spirit to reveal to us what He wants us to know.

When we are in close communion with Him—avoiding sin, reading His Word, fellowshipping with other Christians, and obeying His will—our communication with Him is clear. However, sin can distort this communication and hinder us from hearing God's voice. We need to be vigilant about confessing our sins so we can continue to experience the joy of communicating with Him.

Even when we are sinning, God can communicate with us; however, we can have truly cosmic conversations with Him when we consistently ask Him to forgive us of our sins.

The More We Know Him, the Better We Can Communicate with Him

Communication is always easier when you know the person you're communicating with—and the better you know that person, the deeper your connection will be.

As a business owner, I often receive form letters from people who are trying to convince me to stock their products. They pretend they know me and my business, even though we've never had any type of personal interaction, and they've never visited my store. To me, it's obvious that their letters are mass-produced form letters. For the most part, they go right into the recycle bin because those people don't know me, my business, my situation, my cares and concerns, or my skill set.

Prayer is most effective when we acknowledge that we are actually communicating with the Creator of everything on Earth and throughout the cosmos—the all-powerful One Who created us in His image and loves us more than we could ever imagine. When we recognize just how powerful He is and how much He loves us, it helps us understand His infinite ability to respond to our prayers.

When I read His Word—the Holy Bible—I get a better understanding of His character and characteristics. I don't fully understand many of them. Yet even though there are a lot of events in both the Old Testament and the New Testament that I don't fully comprehend, reading His Word does give me a sense of His love for humanity—His love for me, for you, those we love, and even those who oppose Him and His ways. He loves us, *regardless of where we stand in our faith*, and even though we are sinners:

> “For one will scarcely die for a righteous person—though perhaps for a good person one would dare even to die—God shows his love for us in that while we were still sinners, Christ died for us.”
>
> **—Romans 5:8**

What an expression of love!

How can we do anything but revere, honor, and worship Him and call on Him at every turn, knowing that the Creator of the Andromeda Galaxy sent His only Son, Jesus Christ, to die on the cross that we might be forgiven? He longs for us to be restored to a right relationship with Him. That is why He created us.

Try Journaling Your Prayers

We typically think of prayer as a verbal communication with God, even if we do it silently. However, I have found that sometimes, I can pray more effectively when I journal (write, type, or dictate) my prayers. When I write or type my prayers, it helps me be more intentional about what I want to express.

When I take a moment to reflect on His creations, both great and small, it makes it easier for me to pray. When I dictate my prayers into a recording app on my cell phone or type them into my phone or computer, I pray that God will help me use the electronic device to clarify what I want to say to Him.

Journaling my prayers also enables me to keep a record of what I’ve been praying about when I get stuck. Sometimes, when I’m not sure what to pray about, I will search the internet for a James Webb telescope photograph or an electron microscope photograph, and they remind me of the intensity of God’s majesty and the depth of His love for Me. When I see images of His creations through technology that allows us to view them, I am astounded by God's greatness, and that inspires me to pray more effectively.

Also, when I keep my prayers in a computer file, I can do a keyword search in that file and go back to see which prayers have been answered. When I realize that one of my prayers was answered, that encourages me to keep praying about other things.

Answered prayers encourage us!

I think our prayer life needs to include both formal prayer and informal prayer. We can pray while driving, walking the dog, or preparing to go to sleep, but I believe we also need to set aside a time to focus on doing nothing but engage in prayer with our Creator. Most Christians are comfortable with informal prayers, but many are not as vigilant about engaging in more formal prayer. I believe we need to have devoted prayer time in which we get really honest and raw with God, really pouring our hearts out to Him. I typically engage in formal prayer time with Him in the morning. Each day, I try to take the time to journal my prayers, either electronically or in a written notebook, because it helps me stay focused.

When I speak my prayers into a word-processing program or a voice-recording app, I don't worry about grammar, punctuation, or good syntax. What matters to our Creator, and therefore to me, is the depth of our love for Him, the extent of our commitment to spending time with Him, and the trust we place in Him to lead us to where He wants us to go.

Just like the communication we have with other people, our communication with Him needs to be open, honest, and a regular part of daily life for it to be effective. Our prayers need to be far more than just slogans or form letters. Let us commit to having epic prayers—cosmic conversations with our Creator!

Technology Invented by Humans Helps Us Communicate

Before we had modern printing equipment and technology, it was expensive and time-consuming to communicate messages. Scribes had to write manuscripts by hand. As a result, literacy was low, and Christ followers could not easily spread the Good News throughout the land.

A major breakthrough happened in 1450, when the Gutenberg Press was introduced in Europe by Johannes Gutenberg. It began an information revolution, making it possible to produce and distribute printed materials to the masses in the form of books, newspapers, and other media.[26]

The Gutenberg Bible was the first printed version of the Bible. It consisted of 1,282 pages and was printed in 42-line columns. Experts consider the Gutenberg Bible to have high aesthetic and technical quality However, concerns arose with this capability because information was no longer being controlled by a select few religious and secular authorities.[27]

There are both positive and negative aspects of any new development or invention, from the internet to artificial intelligence (AI). Since the dawn of life, communication has been used to communicate both good and bad messages. The greater our ability to mass-produce information, the greater the chance for misinformation and disinformation to be circulated. We have to be very careful because there is a lot of bad information and influence out there. We have to remain spiritually strong and ignore the bad while using the good to enhance our spiritual lives.

---

26 "The Gutenberg Revolution: How the Printing Press Shaped Humanity and What It Means for AI," QuoCirca, January 22, 2024, https://quocirca.com/content/the-gutenberg-revolution-how-the-printing-press-shaped-humanity-and-what-it-means-for-ai/.
27 Ibid.

Just as the Gutenberg press allowed mass distribution of God's Word for the first time, computers make it easier for us to study the Bible. We can search for words, terms, or Scriptures and study the Bible much more efficiently. When I was in the seminary, a large part of the work we did was to study and search the Scriptures. It was time-consuming and tedious.

For example, if I wanted to find all the mentions of the word "prayer" in the Bible, now I can do a keyword search in any Bible app and find the results within seconds. Before we had this technology, it took hours to do this same type of search manually.

God has empowered us with the capability to develop new technology. As Christians, we need to use these new developments to glorify His Kingdom.

As with just about anything, there are positive and negative aspects of technology. We can use technology such as the internet to increase our knowledge, communicate and fellowship with other Christians, spread God's Word, and study God's Word. However, the internet also provides easy pathways to inappropriate, ungodly, and evil content. I believe it is our duty as Christians to take advantage of the benefits technology provides us while using caution and discernment to avoid the negative aspects—and to keep children safe.

The way to overcome evil is not to ignore it, but rather to counteract it with godly actions.

The Positive and Negative Consequences of Alfred Nobel's Inventions

Here is another example, outside the realm of communication, of how an invention can result in both good and bad consequences.

Alfred Nobel (1833–96) was a scientist, author, pacifist, and inventor who held 355 patents. His father, Immanuel Nobel, had experimented with using nitroglycerine to create explosives, and

Alfred continued this work. However, as they discovered, nitroglycerin is unstable and dangerous. In 1864, the Nobels' nitroglycerin factory blew up, killing Alfred's younger brother, Emil, and several other people.[28]

Nobel continued to experiment with safe ways to detonate explosives, and in 1867, he invented dynamite, which combines nitroglycerin with absorbent material like wood pulp or diatomaceous earth to make it safer to handle. Dynamite became a safer and more versatile alternative to black-powder explosives. When he died in 1996, he left the bulk of his fortune in a trust to establish the Nobel Prizes, which became, and still are, considered the most highly regarded international awards.[29]

Although Nobel's inventions led to many positive advances in construction, wartime defense, and other fields, he was criticized for the destructive nature of his inventions. After his brother died in the explosion, some people began calling Nobel "the merchant of death." Experts think he may have left his fortune to establish the Nobel Prizes as an effort to help outweigh the negative consequences of his inventions with positive outcomes.[30]

He wanted to inspire and encourage people to use their God-given abilities to create developments that make a positive impact on society and the world.

As Christians, we must not let technology intimidate us, nor should we be afraid of it. Instead, let us benefit from embracing technology and learning how to use it to advance God's Kingdom and to empower us in our faith journeys.

---

28 "Alfred Nobel, Swedish Inventor," *Britannica*, last updated March 25, 2025, https://www.britannica.com/biography/Alfred-Nobel.
29 Ibid.
30 Ibid.

A Prayer to Our Holy Father, Expressing Awe at His Creations

Heavenly Holy Creator, our heavenly Father, I am in awe of Your creations. I am amazed at how Your creations go beyond the Earth and out into the cosmos. Thank You, Lord, for empowering humanity to discover and harness technological wonders, such as those that enable us to launch space probes such as *Voyager 1* and *Voyager 2*. I acknowledge, Lord, that You are truly the Creator of all things on Earth and the heavenly realms beyond.

If *Voyager* or another space probe were to encounter other intelligent life forms, then I know, Lord, that You created them, too, and sustain them. You created everything in existence, no matter how far away or how small. This reminds me of Your incredible greatness, which expands beyond our imagination.

I am in awe of the fact that that this Earth I stand on is just one small marble in the vastness of space. Also, even though I am only one human being among the billions on this planet, You know everything about me. You have even numbered the hairs on my head. Your love for all of us is beyond measure. Thank You for creating us all in Your image and for sustaining us.

In Jesus's name I pray. Amen.

Actions I Will Take:

1. Continue using human-made technology to journal my prayers, spread God's Word, and learn more about His awe-inspiring creations.
2. Encourage Christians not to be afraid of technology, but rather to embrace it as a powerful tool that can strengthen their faith, while guarding against the dark side of technology.

Now I encourage you to write your own prayer. Think about what we have discussed in this chapter, and pour out your heart to Him. What do you want to say to Him? What do you want Him to communicate to you?

____

____

____

____

____

____

____

____

____

____

Actions You Will Take:

1. ______________________________
2. ______________________________
3. ______________________________

CHAPTER 5
LETTING HIS PERFECTION DISSOLVE OUR UNCERTAINTY

Real Planetary Nebula NGC 1514 (the Crystal Ball Nebula) in Taurus, taken by a CCD camera through a medium-focal-length telescope.

The photo shown above was taken of a nebula called NGC 1514[31] in 2016 by a CCD camera[32], five years before the more advanced James Webb telescope was available.

---

31 NGC 1514, also known as the Crystal Ball Nebula, is a planetary nebula in the zodiac constellation of Taurus. See "With NASA's Webb, Dying Star's Energetic Display Comes Into Full Focus," Michael Ressler and Dave Jones, NASA, April 14, 2025, https://science.nasa.gov/missions/webb/with-nasas-webb-dying-stars-energetic-display-comes-into-full-focus/.

32 A CCD, or charged-coupled device, is an electronic sensor that converts light to digital signals through charges generated by bouncing photons on a thin silicon wafer. CCDs were the gold standard for camera sensors from the early 1980s until

If you go to the Webb Space Telescope website,[33] you will see two more images of the same nebula. Because these images were taken with more advanced technology than the image above, we can get a much closer look at the nebula. (I cannot publish them in this book because of copyright laws.) On that website, the photo on the left was taken with the WISE telescope, which was invented in 2009, and the one on the right was taken by the James Webb telescope, which was developed 12 years later, in 2021.

As you can see, the image on the right is much clearer. That clarity of detail makes it easier to see the nebula's structure and characteristics.

The discovery of nebulas is like prayer—the more knowledge we gain, the more we understand them.

For example, people often ask me to pray for someone they care about, but I do not know the person or details about his or her situation. I still pray, asking God to help that person, even though I have a limited understanding. God knows what each person needs, even if we don't.

What often happens is, a couple of weeks later, the person who asked me to pray will see me, and I'll ask how the person is doing. My acquaintance will provide me with an update and more details. So then, when I pray for the person at that point, I have more details and can be more specific in my prayers as I ask God to help the person.

Similarly, nebulas have been around for millions of years, even though we couldn't see them—and, for a long time, didn't even

---

the late 2000s. They are still the preferred sensor used in certain areas of photography. See "What Is a CCD (Charge-Coupled Device), and How Is It Used?" Jayric Maning, MakeUseOf, January 9, 2023, https://www.makeuseof.com/what-is-a-ccd-charge-coupled-device/.

33 "Planetary Nebula NGC 1514" (WISE and Webb Images Side by Side), James Webb Telescope website, https://webbtelescope.org/contents/media/images/2025/118/01JRDFFM4H78GZYGBDDDNXZ95Y.

know they existed. But then, over time, technology has enabled us to learn more details about them. As our awareness grows, our knowledge grows.

It's like looking through a backyard telescope first—the details aren't clear. And that's OK. Again, when we lift someone up in prayer, we don't need to know the details about their struggles because God does. In our youth group, we called these unspoken prayer requests simply "praying for someone by name without knowing all the details."

For example, if we are praying for someone who is battling an addiction to drugs or alcohol, we don't know exactly what that person is going through, but we can still pray for God to help him or her because He *does* know what that person needs to heal. We can simply pray for that person to be released from the shackles of addiction: "God, please help *[Name]* as he/she works through his/her trials. Be with him/her, give him/her strength and comfort, and open his/her heart so he/she will cry out to You and trust You to provide help."

Over time, as we get to know people better and learn more about their situations, we can make our prayers more detailed as we ask God to help them.

The World's Most Powerful Telescopes

Using the intelligence, curiosity, and skills that God blessed humans with, scientists have developed several incredibly powerful telescopes. These sophisticated human-made instruments enable us to study God's creations. Here is an overview of three of the most powerful telescopes.

1. **Hubble Space Telescope**—NASA launched the Hubble Space Telescope on April 4th, 1990. From determining the atmospheric composition of planets around other stars to

discovering dark energy, Hubble has changed humanity's understanding of the universe. Its design, technology, and serviceability have made it one of NASA's most transformative observatories.[34]

2. **WISE and NEOWISE**—On December 14, 2009, NASA launched the Wide-field Infrared Survey Explorer (WISE), a highly sensitive astronomical telescope that surveyed the entire sky in four mid-infrared bands spanning from 2.6 to 26 microns. WISE cataloged hundreds of millions of astronomical objects, comprising many asteroids (several hundred of which are located near Earth), brown dwarf stars (including some of the stars closest to the sun), and ultra-luminous infrared galaxies. WISE mapped the entire sky twice. WISE left behind a massive trove of data covering more than 740 million asteroids, stars, galaxies, and other cosmic objects.[35]

   NASA hibernated WISE in 2011 because the telescope ran out of the coolant needed to chill its sensitive infrared detectors. However, two of the spacecraft's four infrared-wavelength bands still worked, and they excelled at finding faint near-Earth asteroids and comets. So, in 2013, after nearly three years of hibernation, NASA woke WISE back up and gave it a new name: NEOWISE (Near-Earth Object Wide-field Infrared Survey Explorer). Its new goal was to aid planetary defense by surveying and studying asteroids and comets that stray close to our region of the solar system.

---

34 "Hubble Space Telescope," NASA, https://science.nasa.gov/mission/hubble/.
35 "NASA's Infrared Survey Telescope Ready to Retire," Caltech, August 8, 2024, https://www.caltech.edu/about/news/nasas-infrared-survey-telescope-ready-to-retire and "WISE: Wide-Field Infrared Survey Explorer," Utah State University Space Dynamics Laboratory, https://www.sdl.usu.edu/downloads/brochures/wise.pdf.

NEOWISE built on WISE's work by identifying and characterizing the population of near-Earth objects (NEOs).[36]

3. **The James Webb Space Telescope**—Three agencies launched the James Webb Space Telescope on December 25th, 2021: the National Aeronautics and Space Administration (NASA), the European Space Agency (ESA), and the Canadian Space Agency (CSA). It does not orbit around the Earth like the Hubble Space Telescope does; it orbits the sun, 1.5 million kilometers (1 million miles) away from the Earth at what is called the second Lagrange point, or L2.[37]

This is a reminder that there are a lot of things we still don't understand about the universe, although over time, our ability to view and explore God's creations has increased exponentially—and will continue to change. Yet even though we know very little about the characteristics of these nebulae, we still know He created them and everything else, both great and small.

So, even when we feel confused about which path to take and when we feel like our communications with Him are cloudy, we need to pray anyway. He created us in His image, and He wants us to communicate with Him all the time.

We Need to Pray Without Casting Judgment

People in this fallen world struggle with all types of sin, addictions, mental illnesses, physical illness and injuries, career struggles, financial hardships, relationship issues, and many other

36 Ibid.
37 "James Webb Space Telescope," NASA, https://science.nasa.gov/mission/webb/.

burdens. *We need to be careful not to judge them, even if we do not agree with the choices they are making.* God tells us that we should not judge others, or He will judge us. In Matthew 7, He reminds us that we are all sinners and have no right to judge one another:

> "Judge not, that you be not judged. For with the judgment you pronounce you will be judged, and with the measure you use it will be measured to you. Why do you see the speck that is in your brother's eye, but do not notice the log that is in your own eye? Or how can you say to your brother, 'Let me take the speck out of your eye,' when there is the log in your own eye? You hypocrite, first take the log out of your own eye, and then you will see clearly to take the speck out of your brother's eye."
>
> **—Matthew 7:1–5**

For example, many people struggle with sexual issues, such as prostitution, promiscuity, and pornography, to name just a few. People get involved in these types of sins as the result of abuse or neglect, and/or poor choices, which leaves them open to believing the lies of the evil one. Many times, people who have been abused become abusers as they grow into adulthood. The sooner we can address abuse in someone's life, the better.

Years ago, when I worked at a hospital, I had a friend who was gay and promiscuous. We'll call him Tom. When other people were around, Tom claimed to be very happy with his lifestyle.

One time, we were working the "graveyard" (overnight) shift together. It was about 2 o'clock in the morning, and it was just the two of us, talking in a quiet area. He knew I was a Christian, and I had always been kind to him.

Tom said, "Chip, I want to tell you something. I know that the way I am living is sinful. Before I die, I want to repent."

Suddenly, several people walked into the area where we were, so he didn't say anything else. Once again, he put on the mask of false happiness. His brief admission gave me insight about how to pray for him. I prayed that God would convict Tom and place the urge in him to repent of his sins and begin living a godly life.

However, in later conversations, I found out that Tom had been molested by his brothers. He knew in the depths of his soul that his lifestyle was not pleasing to God. He wanted to repent but was struggling to do so.

Knowing more about Tom's situation, praying for him to repent was no longer my main focus; instead, I prayed that God would shower my friend with His grace and heal the wounds he experienced as a result of abuse.

Sometimes when we know people are not living godly lives, we might be tempted to pray for them to repent. Is it more compassionate to pray for God to heal them by His grace than it is to just pray for them to repent? Sometimes repentance comes before healing, sometimes after, and sometimes simultaneously.

Many times, we make assumptions about people without knowing anything about them. We all need to avoid doing this.

For example, if I see a panhandler on a street corner with a shopping cart containing everything he owns, I might think, "Why doesn't that lazy man just get a job? It's so annoying that he is standing there, begging for money."

The man could be an alcoholic or a drug addict. What if that man lost his job through no fault of his own, then lost his house, and he is caring for a sick wife or a sick child? What if he's a military veteran who has post-traumatic stress disorder and is unable to work or sign up for government resources?

I don't know his story, but I can still pray for him. I can say a prayer that goes something like this:

Lord, I know that You are a loving God and that Your Son, Jesus, died on the cross for all of us. I can see that this man is really struggling right now. I pray that You will begin the healing process in his life so he can step out of the darkness he's in now and move toward the light—toward You. Even if this person has done something dishonest or unethical, please help him. I pray that he will come to know You. Forgive me, Lord, for being judgmental about him. Please help me be more compassionate—more like Jesus. In Jesus's name I pray. Amen.

There are so many wounded people in this world. In fact, each of us has our own individual set of wounds. And those wounds have impacted each of us differently. I was abused at a summer camp. How did that impact me? I don't know for sure, but sometimes I have trouble allowing people to get close to me. Unhealed wounds—whether physical, emotional, psychological, or otherwise—can influence people's behavior and can lead us toward sin. Sometimes, we just need to pray with conviction that our Creator can heal the wounds, even if we don't know much about how the wounds are affecting us and others.

Sometimes, we aren't even sure what kind of help *we* need. The only prayer we can manage to get out is, "God, please help me!" That is a complete prayer. It is heartfelt and honest. God knows what we need; we don't have to know exactly what will help us in those moments of despair. But He does:

> "Likewise the Spirit helps us in our weakness. For we do not know what to pray for as we ought, but the Spirit himself intercedes for us with groanings too deep for words."
>
> **—Romans 8:26**

God created us. He knew us before we were born. He knows what's in our minds and hearts. He knows the details about why we do what we do, and what we need, even if we don't—just as He knows all the characteristics of all the star nebulae out there, even if we can't see them well with our limited technology at this time.

Our lack of knowledge about the details should never limit us in our prayers because we are praying to the Limitless One—the Creator and Sustainer of the universe.

Prayer During Times of Hardship

Uncertainty is uncomfortable, and it can cause us to stumble in our faith. This is certainly true when we experience hardships; uncertainty is just one of the many painful aspects of the trials and tribulations we face in life. It's difficult to understand why bad things happen to us, even if we are doing our best to serve God.

As I mentioned, in 2014, I began to experience one life-altering hardship after another, and the bad luck continued for three years.

At first, my prayer life faltered because I was angry at God. I couldn't understand why all these devastating things were happening to me. I really struggled to accept my circumstances. However, I didn't want to abandon my faith. I continued to read my Bible and decided to continue going to church, even though I was struggling. After the divorce, I had begun attending a megachurch, with several thousand members, because I wanted to get lost in the crowd. After twenty years of pastoring my own church, I now wanted to just fade into the background. I was caring for two-year-olds during the main Sunday-morning services. I continued to serve in that role while attempting to accept my circumstances.

The new church took up a collection and gave me gift cards to help me get back on my feet, and I was grateful.

I still loved God, even though I was upset with Him. You can love someone and be angry with them. I prayed, but my prayers became less about expressing gratitude to Him and more about expressing my frustration and anger. Eventually, I came to accept what was happening to me. I embraced the Scriptures that remind us that *everything*—both the good and the bad—works together for the good of His Kingdom:

"And we know that for those who love God all things work together for good, for those who are called according to his purpose."

—Romans 8:28

"Count it all joy, my brothers, when you meet trials of various kinds for you know that the testing of your faith produces steadfastness. And let steadfastness have its full effect, that you may be perfect and complete, lacking in nothing."

—James 1:2–4

After that, my prayer life eventually became more intense.

Sometimes, trials and tribulations bring us closer to God. I don't think God *caused* my cancer, my divorce, the fire in which I lost everything, or my brain tumor. People make the mistake of saying God *causes* their hardships, but usually, they are an indirect result of "the fall"—Adam and Eve's original sin in the Garden of

Eden. Satan, the evil one, delights in our hardships. He tries to convince us that our difficult times prove that God does not always love us. We cannot let our discouragement overtake our faith. We must not fall for Satan's lies.

And even though God doesn't usually cause our hardships, He can use them to bring us closer to Him. He also uses our trials and tribulations as a way to help others who might experience similar setbacks. Having gone through them ourselves, we are able to encourage others to stay strong in their faith.

No matter our hardships, we must never let anything separate us from our love of God:

"For I am convinced that neither death nor life, neither angels nor demons, neither the present nor the future, nor any powers, neither height nor depth, nor anything else in all creation, will be able to separate us from the love of God that is in Christ Jesus our Lord."

—Romans 8:38–39

A Prayer to Our Holy Father, to Bring Us Clarity Where There Is Cloudiness and Confusion

Dear Lord Jesus, Creator and Sustainer of heaven and Earth, You created NGC 1514 and all its details long before we even knew it existed and before we saw a glimpse of it through a telescope. In the future, You might give humankind even more sophisticated capabilities to get a closer look at Your magnificent creations. Even though we don't know much about the marvels that are Your creations, we know You made them.

In the same way, even though we might not understand everything about You, lead us to pray to You about everything. Give us a hunger to communicate with You, even if we feel the communication is cloudy.

Lord, I pray for my family members. I pray that You will forgive me for some of the stupid things I've done as an imperfect Dad, husband, son, brother, and uncle that have probably had a negative impact on my loved ones. I pray that You would heal their wounds, even though I don't know the nature, extent, or impact my actions have contributed to their suffering. I pray that You would lead them to make wise decisions. May Your grace work in their lives so they will feel led to seek out Your wisdom and guidance.

Lord, I pray that You will bring clarity to them as they live their lives. Lead them to acknowledge Your majesty and magnificence. Give them strength to resist the evil one's lies. May Your grace transcend the lies and the noise in this fallen world we live in. Please open their minds and quicken their hearts to Your truth.

In Jesus' name I pray. Amen.

Actions I Will Take:

1. Find a way to impact young parents in a positive way.
2. Be more sensitive to those around me.
3. Be more open and honest in my relationships.

Now I encourage you to write your own prayer. Think about what we have discussed in this chapter, and pour out your heart to Him. What do you want to say to Him? What do you want Him to communicate to you?

Actions You Will Take:

1. ______
2. ______
3. ______

CHAPTER 6
COMPARING THE VASTNESS OF SPACE WITH HIS ABILITY TO HEAL

In this photo of the Earth taken by the *Apollo 17* moon mission crew, the view extends from the Mediterranean Sea to Antarctica.

The crew of *Apollo 17* took the now-classic "Blue Marble" photo on December 7, 1972, 18,300 miles from the Earth's surface.[38]

---

38 "The Blue Marble: The View From *Apollo 17*," NASA, April 20, 2020, https://www.nasa.gov/image-article/blue-marble-view-from-apollo-17/.

It is a striking photograph, as is the date—December 7th, known as Pearl Harbor Day—because on December 7, 1941, the Japanese launched a surprise aerial attack on the US naval base at Pearl Harbor on Oahu Island, Hawaii, precipitating the entry of the United States into World War II.[39] December 7th is still recognized as "a day that will live in infamy," in the words of President Franklin D. Roosevelt during his famous speech on December 8, 1941.[40]

These two historical events occurred thirty-one years apart. One shows the greatness of human potential to build and discover. The second is an example of our ability to hate and destroy. What surpasses both factors is God's vast, all-encompassing love and capacity to heal.

NASA's *Apollo 17* was the last mission to put a man on the moon. It marked the end of six missions over six years with twelve men walking, riding, standing, leaping, and even playing golf on the lunar surface.

In contrast, Pearl Harbor ushered the United States into World War II. The attack killed 2,403 US personnel, including 68 civilians, and destroyed or damaged 19 US Navy ships, including eight battleships. Japan surrendered only after the United States dropped atomic bombs on Hiroshima and Nagasaki.

As you can see, December 7th is both a day of accomplishment and infamy. Achievement and adversity. Great glory and horrible tragedy.

We have learned so much from science and technology, yet we are still fallen. Science and technological advances alone will not solve the broken human condition. The evil one still prowls about,

---

39 "Pearl Harbor Attack," *Britannica*, last updated April 1, 2024, https://www.britannica.com/event/Pearl-Harbor-attack.

40 "FDR's 'Day of Infamy' Speech: Crafting a Call to Arms," *Prologue Magazine 33*, no. 4 (Winter 2001), https://www.archives.gov/publications/prologue/2001/winter/crafting-day-of-infamy-speech.html.

lying in an effort to generate hatred, wars, suffering, disease, racism, and many other struggles.

The humans inhabiting this beautiful blue marble rotating in space are still fallen. There is "no one righteous, not one" (Rom. 3:10). This is why The Holy Spirit inspired Paul to write the two-word verse "Pray constantly" (1 Thess. 5:17, HCSB). He must have known what the future might hold.

Another famous photograph from the Apollo program is called "Earthrise." I encourage you to take a look at this image on NASA's website.[41]

Taken on Christmas Eve 1968 by Bill Anders as part of the *Apollo 8* mission, it shows the Earth rising above the desolate moonscape. This photo became a powerful image credited with helping kick off the environmental movement. Although 1968 was a tumultuous year of unrest, protest, and violence, this photo helped close the year on a positive note. The contrast of the blue marble against the dark void of space is beautiful. There is a striking contrast between Earth in all its glory viewed from the desolation of the lunar surface.

---

41 You can see the iconic "Earthrise" photo on NASA's website at https://www.nasa.gov/image-article/apollo-8-earthrise/.

This image of the Earth rising above the moon's surface is very similar to the Earthrise photo.

This image reminds me that the darkness and desolation surrounding our human condition will never dim His light and love. His grace, glory, redemption, and restoration will always penetrate the darkness. My hope is that images like these will inspire us to hope, to pray, and to work diligently for a better tomorrow.

Often, I go out into a remote area with my telescope to gaze at His creations. And sometimes, it's scary. I can hear coyotes howling from every direction. But then I realize God is with me, and there is no reason for me to be afraid of coyotes. There I am, looking at the vast universe that He created, yet momentarily, I'm worried about coyotes. The powerful God who created everything can certainly protect me from a coyote!

Sometimes, I hesitate to pray to God. I think, "He has more important things to do than to listen to my prayers." Other people have shared with me that they feel this way, too. Yet none of our prayers are silly or insignificant to Him. Satan puts that thought into our minds in an attempt to keep us from praying to Him.

When I was a minister, we regularly held a time of sharing and prayer requests. People would ask for the congregation and me to pray for them or someone they knew. Sometimes, the prayers were extremely serious. For example, someone might say, "Please pray for my mom. She was just diagnosed with a very aggressive form of breast cancer. She's undergoing chemo and radiation, and the doctors aren't sure of her prognosis."

I think everyone would agree that a prayer like this is important.

Many times, after someone in the congregation made a serious prayer request like that, someone else would say, "Please pray for me because my car broke down."

This annoyed me at times! I would think, "This other man just told us his mother is dying of cancer, and you are worried about your car?!"

Again, that was Satan, trying to minimize the importance of people's prayers in an effort to keep us from praying to our Creator.

Every one of our prayers matters to God. Never feel like God is too busy to hear your prayers! He wants us to take everything to Him in prayer.

See the True Power of the One We Pray To

As a biblical theologian with some scientific training, I believe theology and science are not enemies of each other, but disciplines that complement one another. God created us with the capacity to explore His universe. Sometimes we sell God short. We limit our acknowledgment of His creation of the Earth and all that is on this planet.

God is all-powerful. There is nothing greater than Jehovah. All things were created by Him and for Him. This includes not just the things we can see, but also the things we cannot see. The forces that hold the universe together. Planets as well as gravity. Elements as well as protons, electrons, and the forces between them. God created molecules and the ionic, covalent, and metallic bonds holding molecules together.

His creation did not start with Earth. It started with light—photons that behave both like particles and waves. Physics, biology, mathematics, and chemistry are some of the disciplines humans developed to discover and explain the ways and means of God's creation. *Theology* is a study of the nature of the Divine, the One who created what science studies.

This book is not meant to convince you of the existence of God. Nor is an attempt to convince you of the truth of creation, Christianity, or any particular branch of theology (if that happens, though, great!). Instead, it is intended to empower and encourage Christ followers to pray more diligently, to see the true power within the One we pray to.

A Prayer to Our Holy Father, to Understand Human Suffering

> Jehovah Jireh[42], as I look at photos of the Earth taken from space, such as the one of the Crystal Ball Nebula in chapter 5, I am reminded of Your creative power, as well as Your love for all the people on this planet. At the same time, Lord, I am reminded that we whom You created in Your image are still fallen. Imperfect. We have the capacity to do great good as well

42 "Jehovah Jireh" is one of the Hebrew names for God. Its first appearance is in Genesis 22:14 (KJV), written as one word, "Jehovahjireh": "And Abraham called the name of that place Jehovahjireh: as it is said to this day, In the mount of the Lord it shall be seen." Read Genesis 22 for context.

as the capacity to destroy. To love or to hate. To work toward peace or to seek or even embrace conflict.

Sometimes I struggle with our capacity to do so much harm to so many people, as well as to Your creation itself. There is so much suffering in this world, Lord. I know all suffering is either directly or indirectly a result of our fallenness and the choices we and/or others make. Sometimes, Lord, I choose to step outside Your will. I choose the path of darkness over the light of Your love. Forgive me for these choices.

I am so thankful for the freedom You have given us—free will—to make our own choices. I am thankful that we're not perfectly programmed robots but instead have the freedom to follow You or not.

Father, I'm often confused and struggle with the amount of suffering and injustice in this world.

I pray that my struggles and lack of understanding do not lead me or others astray. Strengthen our faith, Lord by Your grace, when we doubt. Even in in our suffering, empower us to follow You. I pray that You give a spirit of understanding to Your followers in this present day so we can understand why there is so much suffering in the world. I pray that all men and women of faith develop an understanding of the theology of suffering. I pray every sentence in this book encourages believers to stay strong, no matter what comes their way.

Jehovah Jireh, I often have conversations, debates, and even arguments with people I love. They often deny Your existence because of what they view as injustice and suffering. Suffering is such a universal component of living on this planet. I know, Lord, that in many instances, people in Your Church have inadvertently inflicted injustice and suffering on others in

Your name. Sometimes, we allow our own self-righteousness to cloud our judgment, and we say things that are hurtful rather than helpful. As a result, we hurt others—often inadvertently.

Forgive us, Lord, for the times we have harmed others. Forgive us for our mishandled attempts to convince people of Your love and righteousness. Help us serve as positive ambassadors of Jesus Christ so that our words and actions will encourage our Christian brothers and sisters to look to You for understanding. Help us remember Romans 8:28: "And we know that all things work together for good to them that love God, to them who are the called according to his purpose."

May our lack of understanding about suffering not drive us *away from* You, Lord, but *toward* You so we can gain a deeper comprehension.

In Jesus's name I pray. Amen.

Actions I Will Take:

1. Learn to be aware of, and to curb, my own self-righteousness. Be more sensitive to the Holy Spirit and allow the Holy Spirit to dissolve my own self-righteousness.
2. Think about the words I am going to say and ask myself, "Is this true? Is it necessary? Is it helpful?" If what I am planning to say isn't all three, I will keep it to myself.

Now I encourage you to write your own prayer. Think about what we have discussed in this chapter, and pour out your heart to Him. What do you want to say to Him? What do you want Him to communicate to you?

Actions You Will Take:

1. ______________________________
2. ______________________________
3. ______________________________

CHAPTER 7
SEEING THE CREATIVITY IN COMPLEXITY

God created the universe and everything in it—from the smallest cells to the largest galaxies. Let's explore and contrast some of God's smallest and largest creations, to further understand and appreciate His power.

The Complexity of God's Smallest Creations

The image shown below is an *astrocyte*, which is a subtype of glial cells that make up the majority of cells in the human central nervous system. Astrocytes perform metabolic, structural, homeostatic, and neuroprotective tasks such as clearing excess neurotransmitters, stabilizing and regulating the blood-brain barrier, and promoting synapse formation found in the brains of mammals.[43]

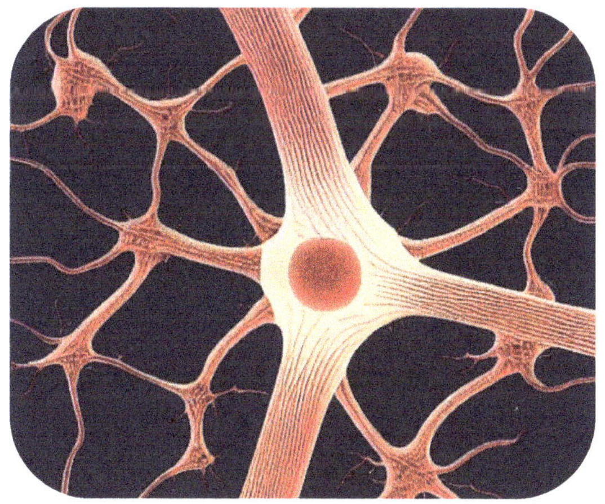

An astrocyte.

---

43 "Histology, Astrocytes," Dereck C. Wei and Elizabeth H. Morrison, National Library of Medicine, last updated May 1, 2023, https://www.ncbi.nlm.nih.gov/books/NBK545142/.

Astrocytes are so small that we cannot see them with the naked eye. Yet they are responsible for many complex functions in the CNS, including regeneration of CNS damage, transmitting signals between neurons, maintaining pH and fluid levels in the brain, removing free radicals from the CNS, and coordinating cerebral blood flow, to name just a few.[44]

Want to know something interesting? I began writing this chapter before I found out I had a brain tumor.

After my surgery, I reviewed what I had written again. I was stunned to realize that the astrocytes I had written about most likely played a huge role in my own recovery! When my tumor was removed, the astrocytes in my brain were overwhelmed and likely had to regenerate new pathways to heal my central nervous system.

The tiny neurons and synapses in the human brain are just as complex as the vast galaxies in the universe—and God created all of them.

An unusual experiment conducted at the Allen Institute for Brain Science in Seattle shows just how complicated our brains are. It reveals how brain cells connect to form what we experience.

A Baylor College of Medicine team genetically engineered a mouse so that the neurons in its brain would glow when active. They showed the mouse a variety of video clips featuring sci-fi, nature scenes, animations, and sports. Then they used a laser-powered microscope to capture the activity in the visual cortex of its brain. They "shaved" the brain sample into more than 25,000 slices, each one thinner than a human hair. they used an electron microscope to capture almost 100 million images. Those images revealed the

44 Ibid.

fibers—axons and dendrites—that connect the neurons, like spaghetti in a bowl.[45]

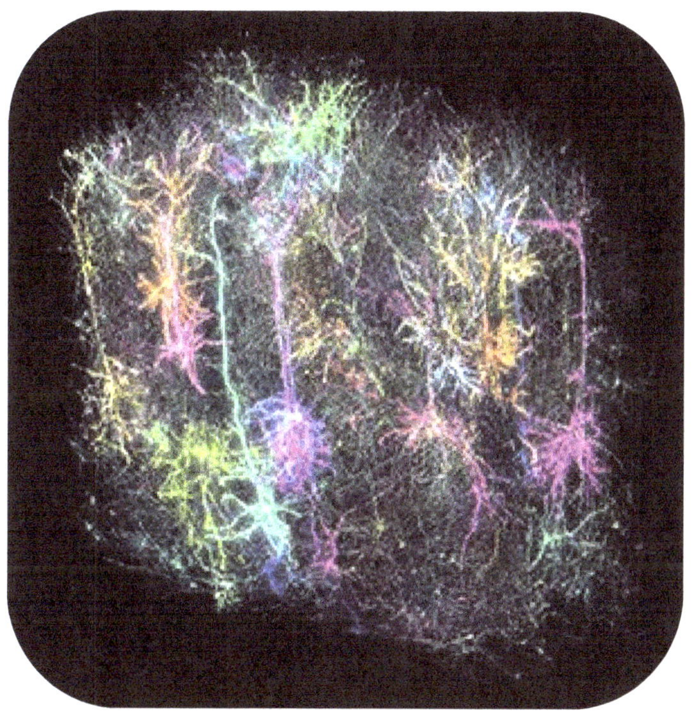

This image is a digital representation of neurons in a section of a mouse's brain. From the Allen Institute for Brain Science. Used with permission.

Next, scientists used powerful software and artificial intelligence to stitch the images back together. Their digital reconstruction recreated the tangled circuits as 3D images. Each neural fiber was given a different color, which enabled researchers to distinguish the fibers with clarity.[46]

---

45 "Huge Advance in Brain Research that Was Supposed to Be 'Impossible'," Eric Ralls, earth.com, April 15, 2025, https://www.earth.com/news/brain-mapping-research-mouse-neurons-revolutionary-advance/.
46 Ibid.

This kind of brain mapping holds a lot of promise for scientists to discover how to cure mental-health illnesses and other disorders. It also reminds me of God's great power and creativity.

One of the project's leading researchers commented on this incredible research, saying, "It definitely inspires a sense of awe, just like looking at pictures of the galaxies. You get a sense of how complicated you are."[47]

The Complexity of God's Largest Creations

Now let's look at some of God's massive creations. They, too, are incredibly complex.

The image below is of the Horsehead Nebula—basically a giant cloud of dust and gas located in the constellation Orion. It is approximately 422 parsecs, or 1,375 light-years, away from the Earth. To realize just how big it is, consider that 1 light-year is the distance that light travels in one year, which is about 5.88 trillion miles. This means the Horsehead Nebula is about 8.08 quadrillion miles away from us. That's a number with 15 zeros after it! As for how big the Horsehead Nebula is, it's estimated to be 3 to 4 light-years in size, which means it is about 17.64 to 23.52 trillion miles across. [48]

47 Ibid.

48 "The Horsehead Nebula Is 66 Trillion Kilometers," Scale of Universe website, https://scaleofuniverse.com/en/universe/horsehead-nebula.

The Horsehead Nebula

Talk about extremes!

The small size of the astrocyte and the massive size of the nebula are both hard to fathom, much less to compare and contrast. Both require human-developed instruments for us to see them. Both contain mysteries not yet fully understood. Both are subject to the laws of physics and chemistry. More importantly, both were created by El Elyon, El Shaddai. These are two of God's titles or names found in the Old Testament, meaning The Most High God and Lord God Almighty.

Looking at these scientific images in the context of the theological truth of creation reminds us how great God is. He is great enough to create something so small and also something so large, simply with His words.

When we pray, we converse with The One whose power and might span the distance between these two types of creations, as well as all the creations in-between and beyond.

It is impossible to describe fully the idea of an astrocyte or a nebula. Both of these creations are delicate and powerful in their own way, and each one has its purpose and place in the universe. Both were created and are sustained by their Creator—the same being

Who created us and Who sent His Son, Jesus, to Earth on our behalf. This is Jehovah Jireh, who hears and responds to our prayers.

It took more than 1,300 years for the light of the Horsehead Nebula to reach our eyes, exposing not only its features but also God's power to create all that there is, was, and ever shall be.

When we pray, we tap into that power!

It is mind-blowing to me that the same God who created astrocytes and galaxies, along with the rest of the universe—this powerful, almighty Creator—loves us, hears our prayers, and sent His Son to die for us so we might have everlasting life.

A Prayer to Our Holy Father, to Awaken Faith in My Unbelieving Loved Ones

> Holy heavenly Father, Creator of the great and small—the giant nebulae, the galaxies, and the planets, as well as the microscopic astrocytes—I come to You in humble amazement and adoration.
>
> You are the all-powerful Creator, and I thank You that I am created in Your image and that, even though I am fallen, Your love for me and all people spans the extremes. Your grace crosses the chasm between our fallenness and Your holiness through Your Son, Jesus Christ, and the ultimate sacrifice He made for us on the cross.
>
> I pray, Lord, that Your grace will awaken the faith of my family, friends, and neighbors who do not yet believe. Draw them near to You to embrace what Your Son did on the cross. Deliver them from their fallenness, and reveal to them, Lord, Your great love for them. Whether it is in a quiet or dramatic way, Thy will be done.

I ask, Lord, that You heal the hurt I have inflicted on those I love. Make them whole by Your grace. Forgive me, Father, for the fallenness that sometimes displays itself in my actions. Thank You that there is no hurt You cannot heal, no sin that You cannot forgive. Thank You for Your greatness, Lord, that You truly span all the extremes. May Thy will be done.

In Jesus's name I pray. Amen.

Action I Will Take:

1. Seek forgiveness from people I have harmed or slighted in any way.

Now I encourage you to write your own prayer. Think about what we have discussed in this chapter, and pour out your heart to Him. What do you want to say to Him? What do you want Him to communicate to you?

Actions You Will Take:

1. ______________________________________
2. ______________________________________
3. ______________________________________

CHAPTER 8
HEALING OUR SOULS WITH MUSIC AND PRAYER

One of the reasons I wrote this book springs from one of the items on my "bucket list." For the past couple of years, I have been learning to play the violin. The instrument creates beautiful sounds when played properly—and it makes the most annoying sounds when handled poorly (just ask my dog).

Throughout the Bible, God mentions music as a way to praise and worship Him. In fact, most of the psalms in the Bible are songs of prayer and praise. Music is just one way we can express the joy of our salvation.

Praise him with trumpet sound; praise him with lute and harp! Praise him with tambourine and dance; praise him with strings and pipe!

—Psalm 150:3–4

The Healing Power of Music

The right kind of music can comfort us. The Bible describes how music that David played on a lyre, a stringed instrument, comforted Saul during his darkest moments:

"And whenever the harmful spirit from God was upon Saul, David took the lyre and played it with his hand. So Saul was refreshed and was well, and the harmful spirit departed from him."

—1 Samuel 16:23

Music is one of the most wonderful blessings of life! In fact, listening to music has many profound benefits to our mental and physical health.

Studies show that music offers comfort and solace in times of distress. It can provide a sense of connection, support, and understanding during difficult times. The history of music in health care has shown that music can reduce stress, anxiety, depression, and even chronic physical pain. Scientists have also found that listening to music helps release *endorphins*—hormones associated with pleasure (and reducing pain), which further reduces stress levels and may even aid in the healing of both physical and emotional wounds.[49]

Listening to uplifting songs and singing along with them can help boost our moods by increasing serotonin in the brain—the neurotransmitter responsible for regulating our emotions and feelings of happiness or sadness. Listening to music can also increase the release of dopamine, a neurotransmitter associated with pleasure and reward. This release can lead us to feel happiness and enjoyment. Music can even improve cognitive function by stimulating areas in the brain related to memory recall. This can help one learn or retain new skills faster than usual. Studies have shown that music can enhance connectivity between the auditory and emotional regions of the brain, which can facilitate memory encoding and recall.[50]

Harvard Health is just one of many sources reporting that the use of music interventions (listening to music, singing, and music therapy) can create significant improvements in mental health.[51]

Psychologists, musicologists, and music therapists also say music can improve our productivity.

So, what kind of music will benefit us the most? Years ago, some studies revealed that classical music was the most conducive to concentration while studying or working. But that line of thought

---

49 "How Music Helps People Heal: The Therapeutic Power of Music," David Victor, Harmony & Healing, June 19, 2024, https://www.harmonyandhealing.org/how-music-helps-people-heal/#:~:text=Music%20helps%20people%20heal%20on,end%2Dof%2Dlife%20situations.

50 Ibid.

51 "Neuroscience Says Music Is an Emotion Regulation Machine. Here's What to Play for Happiness, Productivity, or Deep Thinking," Jessica Stillman, *Inc.*, April 16, 2025, https://www.inc.com/jessica-stillman/the-best-music-for-happiness-productivity-studying/91173726?utm_source=newsletters&utm_medium=email&utm_campaign=INC%20-%20This%20Morning%20Newsletter.2025-04-17%20-%207848&leadId=22996283&mkt_tok=NjEwLUxFRS04NzIAAAGZ4u1Pu7olaEvfzXmImBih56kilWgs0-6exRVc5uCStHZSBqtOKYn6GdeuP2qy5wPs-V5jjYTj6NmG_g3tzW2oH8dxPP2bCaiNQdeoca0Ukyrq.

might be outdated; some researchers say listening to the type of music you personally like most will boost your creativity more than any other.

A Dutch neurologist named Jacob Jolij conducted a study and found that the most cheerful music was up-tempo (between 140 and 150 beats per minute on average), written in a major key, and either about happy events or complete nonsense. In contrast, a psychologist named Emma Gray found that music in the range of 50 to 80 beats per minute can help induce the "alpha state," where the mind is calm but alert, imagination is stimulated, and concentration is heightened.[52]

To find out the tempo of a song, you can count the number of beats per minute (BPM) in the song. Or you can use an app such as Live Tempo from The Apple Store to automatically detect the tempo of a song.

The Fascinating, Complex Way Music Is Created

Think about all that's involved in playing an instrument. There are seven essential musical elements: sound, rhythm, tempo, dynamics, melody, harmony, and texture. They act together to create the powerful and emotive aural phenomenon we know as music.

Physics explains the way vibrations of a stringed instrument produce sound that fluctuates with the frequency of the vibrations. The vibrations can change with the force and speed of a bow as it is drawn across a violin string. Music flows through the air in waves to reach the ears. Different sounds have different wavelengths.

This excerpt from an article titled "The Science of Music" describes the intricate and complex processes that combine to enable us to play and to hear music:[53]

---

52 Ibid.

53 "The Science of Music," Research Features by Karger Publishers, January 18, 2023, https://researchfeatures.com/science-music/.

Sound is created by the vibration of air particles and moves as a wave incredibly fast—the speed of sound is about 343 meters per second. The movement of the air particles begins at the source of the sound—for example, the vibration of a singer's vocal chords, the air blown through a musical instrument, or the vibration of a speaker. When the acoustic wave reaches a person's eardrum, the vibrations are converted to electrical signals. These travel through the body's nervous system and into the brain, which processes the signals, turning them into what we hear as music.

The oscillation or frequency of the sound wave (how frequently the wave repeats itself, measured in Hertz) controls the pitch of the sound, creating low notes and high notes the brain automatically processes. When different notes are played or sung, the air vibrates at a specific frequency. This is very important in musical instrument development; two pianos, when playing the same note, must create a sound wave at the same frequency so the notes sound the same.

This article mentions *oscillation.* We can actually see sound waves when a microphone is connected to an oscilloscope. The microphone changes the sound waves into an electrical signal, and then the oscilloscope shows what the electrical waves look like.

It would be wonderful to see a visual representation of what a note of music looks like when being played. Unfortunately, though, it isn't possible (yet, at least) to photograph sound itself. We can, however, use a photographic device called Schlieren Flow

Visualization to see sound waves. This is possible because sound waves distort the air density around them.

The following are just two examples of countless images of oscillation that photographers have captured:

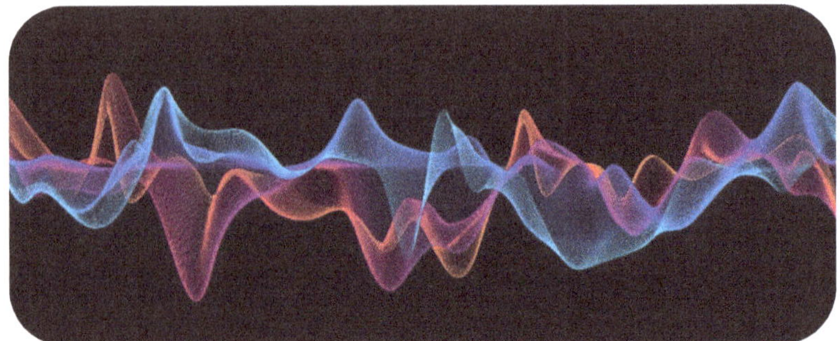

3-D audio soundwave; colorful music pulse oscillation; glowing impulse pattern.

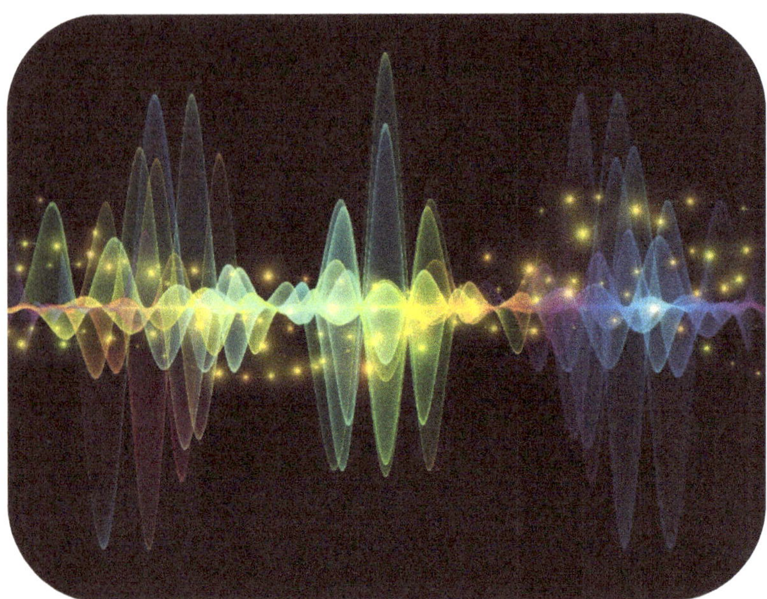

A wave function series. Artistic abstraction composed of colored sine vibrations, light, and fractal elements on the subject sound equalizer, music spectrum, quantum probability, and science education.

Music also involves mathematics. A music instructor once told me, "If you aren't counting, are you really playing music?"[54] Creating good music isn't possible without some basic math. A knowledge of fractions and numeric patterns is essential to understanding tempo, form, chord composition, harmonics, and rhythm. There are musical applications for both algebra and number theory.

In addition to that, imagine all the biological responses and chemical reactions involved when just one musician plays just one note. The contraction, relaxation, and coordination of many muscles at once is critical to moving the bow and pressing the strings. Millions of biochemical reactions take place in the brain, the ears, the eyes, the muscles, the heart, and even the skin to initiate and complete complex processes. The occipital region of the brain "sees" what's on the page. The auditory area hears the notes as they are played. The brain's motor function must communicate with the muscles in the arms, hands, and fingers to make the correct movements.

I asked AI, "How many biochemical reactions does it take to process a single thought?"

Part of the answer was: "A single thought easily involves on the order of 10^{14}–10^{17} biochemical events (hundreds of trillions to a hundred quadrillion+), depending on how big/long you count the "thought." There's no precise cutoff—the point is that cognition rides on an immense, parallel storm of molecular reactions."

Next, I asked AI, "How many biochemical reactions does it take for a single musician to play a single note on an instrument?"

The robot spat out some calculations, but the bottom line it gave is similar to the answer to my previous question about a single thought: "A single note typically requires ~10^{14}–10^{17} discrete biochemical reactions (ATP hydrolyses + other molecular events),

54 Anna S. of Anna's Strings Studio.

with the lower numbers for a tiny, fast finger movement and the higher numbers for loud, breath-driven or sustained notes.

Now multiply those hundreds of trillions of reactions by the number of notes played in a single song and by the number of people in an orchestra—maybe 100. Each musician experiences all those brain-to-body communications to play his or her instrument. Also, the conductor plays a critical role in setting the tempo, keeping the orchestra in time, and ensuring that the orchestra creates a unified, cohesive sound.

An orchestra is divided into four families of instruments: strings, percussion, woodwinds, and brass. Guiding the musicians is a conductor who makes sure all the instruments play in step with the musical composition.

A modern, full-scale symphony orchestra consists of approximately 100 permanent musicians, most often distributed as follows:[55]

Strings:	16–18 first violins, 16 second violins, 12 violas, 12 cellos, and 8 double basses
Woodwinds:	4 flutes (one with piccolo as a specialty), 4 oboes (one with English horn as a specialty), 4 clarinets (one with bass clarinet as a specialty, another specializing in high clarinets), and 4 bassoons (one with double bassoon as a specialty). In theory, all woodwind players are expected to be able to play all auxiliary instruments in addition to their main instrument.
Brass:	4 trumpets, 4 trombones (one with bass trombone as a specialty), and 1 tuba

55 "Orchestra Size and Setting," The Idiomatic Orchestra, https://theidiomaticorchestra.net/14-orchestra-size-and-setting/.

Percussion:	3–4 percussionists (of whom at least one must also play kettledrum)
Other:	1–2 harps and 1 keyboard player (piano, celesta, harpsichord, etc.)

Music acoustics is the study of how sound is produced, transmitted, and perceived in music. It's a multidisciplinary field that combines physics, physiology, and other areas of study. People who specialize in this area of study explore how instruments create sound, how sound travels through different mediums, how the human ear receives sound, the physics of instruments, and the physics of the human ear.

Professional musicians spend decades perfecting their craft. By God's design, all that knowledge and talent is stored biochemically in their brains. The ability to read music, the desire to practice and improve, the ability to master an instrument, and the ability to follow a conductor's prompts are biochemically stored in the billions of neurons that make up the cerebral cortex. Yet this description doesn't even begin to cover all the biological, chemical, mathematical, and physical events that take place in the fraction of a second it takes just one musician to play just one note on the string of a violin.

As we ponder the scientific and theological truths together, we can get a slightly better understanding of the One who hears our prayers. When we pray, we lift our voices to the One who created all that is involved in thinking, speaking, remembering, hearing, seeing, and moving when playing music. We can't help but be in awe of the power and creativity of the One True God when we contemplate the truth of this Creator wanting to hear our prayers. We cannot help but be encouraged to pray without ceasing.

Our ability to play, hear, and understand musical notes is created and sustained by Jehovah Jirah (which means "God provides”). What an awesome, powerful, loving God we pray to in Jesus’s name!

A Prayer to Our Holy Father, to Keep the Church Staff Working in Harmony

Heavenly Father, Lord, as I reflect on music and how complicated, complex, and amazing are the biochemical processes required for us to play, hear, and respond to music. I am amazed by Your creative power at the cellular level. You make it possible for us to play, enjoy, and respond to the many types of music You created. You designed our brains so that we can gain comfort and encouragement from the right types of music.

As I reflect on the other chapters of this book about the complexities of the universe, as well as the complexities of the human body, I see that You, Lord, are the great Designer, Conductor, and Sustainer of this orchestra that makes up the cosmos, from the smallest particle of an atom to the largest black hole. You designed every creation to work in concert with one another. As men and women of faith—people who believe in Your Son, Jesus Christ—lead us to recognize the magnificence of these creations.

At the same time, Lord, I am reminded of how You compared the church to the human body, with everyone in the church having his or her own set of skills, gifts, and abilities. You call pastors to be the conductors of the body to glorify Your Kingdom.

I pray for the pastor of my church, Father. There are a lot of struggles and stresses in that role of leadership. Please surround him and his family with Your protection because I know the evil one wants them to fail. I stand in the gap with prayer against the evil one and his schemes. Help us stand fast in our faith and never compromise; to be compassionate, caring, and loving; and to always be truthful with His beloved children and their families.

Also, I pray for the assistant pastor, youth pastor, music director, small-group facilitators, those who lead the children's ministry, the security officers, the volunteers, those who clean the church, and everyone else who works to keep the many intricate parts of the church working together in harmony so it functions well. May we all function as an orchestra, working in harmony to spread the Good News of Your love to our community and the world.*

In Jesus's name I pray. Amen.

*In the last paragraph of this prayer, I have included people's titles; feel free to use the names of the people in your church, to make your prayers more personal.

Action I Will Take:

1. When I play the violin, I will do so as praise to Jehovah Jirah and not complain that it is a chore to improve my playing ability.

Now I encourage you to write your own prayer. Think about what we have discussed in this chapter, and pour out your heart to Him. What do you want to say to Him? What do you want Him to communicate to you?

Actions You Will Take:

1. ____
2. ____
3. ____

CHAPTER 9
MARVELING AT HIS MIRACLES

A visual depiction of Jesus walking on water, as described in the Gospel of Matthew.

God sometimes suspends the physical properties of His creation to accomplish supernatural events that align with His will. In the Bible, such drastic measures are called "miracles."

For example, the Bible tells us in the Gospels of Matthew, Mark, and John that Jesus walked on water—a physical impossibility unless God suspended His natural law. Matthew 14:22–23 describes Jesus walking on water toward His disciples during a storm at sea. Jesus tells them, "Take heart! It is I. Don't be afraid" (Matt. 14:27). When the disciples first saw Jesus walking on water, they were troubled and afraid because it made no logical sense whatsoever!

In that Scripture, Peter also walks on water, but he begins to sink, and Jesus rescues him.

Walking on water defies two scientific laws: the law of *buoyancy* and the law of *surface tension.*

Buoyancy is the tendency of an object (or a person!) to float or to rise in a fluid (either liquid or gas) when submerged. It's the upward force that a fluid exerts, opposing the object's weight. *Surface tension* is the property of the surface of a liquid that allows it to resist an external force, due to the cohesive nature of its molecules. Jesus had to suspend both scientific laws to walk on water.

We are able to *float* on water because floating is the result of buoyancy. However, people cannot *walk* on water partly because our weight would break the water's surface tension.

Those are the characteristics that God assigned to water when He created it. Yet because Jesus, the Son of God, is all-powerful, He was able to suspend those characteristics for a moment in time to achieve His desired outcome. Jesus suspended the "laws" of science as He walked on water. He also suspended them as Peter walked on water—until fear overcame Peter.

This story show us that when we have faith in Jesus, the Holy Son of our Creator, we can endure any situation, no matter how devastating it seems. It demonstrates His divine power over nature and his ability to be in control, even in the midst of chaos. We can trust Jesus to guide us through life's storms, even when the challenges seem insurmountable—and even when the outcome may not be what we want; it is always best in the eyes of God.

We can trust Him enough to pray about *everything*.

We do not bring people to accept Christ as their Savior and Lord by beating them over the head with the Bible or berating them for their disbelief. Jesus is the only One who can, or should attempt to, do that. We have to use a more common, subtle approach—such as sharing our testimony with our loved ones and praying that God will open their hearts to receive our messages, understand our words, and then accept Jesus as He truly is.

Three Examples of How People Come to Accept Christ: From Subtle to Dramatic

Most people come to trust Jesus Christ as their Savior and Lord in a subtle way—often by praying to God after hearing a sermon in church or hearing someone share their testimony. However, once in a while, these conversions to Christianity are much more dramatic.

It's important to note that it doesn't matter how subtle or dramatic your conversion is; the blessing is the same. If you have accepted Him into your heart, you will spend eternal life with Him in heaven!

I want to share with you the story of three different situations in which God communicated to people His love and mercy. The first story, about Elijah, is from the Bible. He had a very subtle conversion experience that pulled Him toward God, prompted by the "still, small voice" of God.

The second story is about John Wesley, a theologian who lived in the 1700s. He had more of a "supernatural" conversion experience.

The third story is also from the Bible, and it's the most dramatic of the three. It's about how Saul, who spent his time hunting, persecuting, and killing Christians, came to Jesus in a much more dramatic way.

1. Elijah Hears the Still, Small Voice of God

In the Scripture, Elijah experiences a triumph on Mt. Carmel over the worshipers of Baal. Unfortunately, though, it does not result in a revival of true religion in Israel; instead, it leads to a threat on Elijah's life.

After King Ahab's evil queen, Jezebel, orders Elijah to be arrested, Elijah is afraid. He flees and asks for death. An angel appears and offers him food and water to restore him, which enables him to travel to "the mount of God." Then God asks, "What are you doing here, Elijah?"

Elijah begins to complain to God that Jezebel had killed all His prophets, and Elijah felt alone, as the only one who had survived. God instructed Elijah to stand on the mountain in His presence. And then the Lord sent a mighty wind that was so strong that it broke the rocks in pieces, followed by an earthquake and a fire. However, God's voice was not in any of those dramatic displays. Instead, the Lord spoke to Elijah in a "still, small voice"—a gentle whisper.

"And behold, the Lord passed by, and a great and strong wind tore into the mountains and broke the rocks in pieces before the Lord, but the Lord was not in the wind; and after the wind an earthquake, but the Lord was not in the earthquake; and after the earthquake a fire, but the Lord was not in the fire; and after the fire a still small voice."

—1 Kings 19:11–12, NKJV

However, Elijah did not see or hear God in the powerful, natural events. The wind blew ferociously, but Elijah did not sense God. The Earth shook violently, but Elijah didn't feel God. A fire blazed, but Elijah still did not experience God's presence. These powerful occurrences certainly got Elijah's attention, but he did not experience God until he heard God's small, still voice.

This Scripture shows us that God's presence and communications are not always dramatic. Not everyone experiences an earth-shattering, dramatic conversion. Most conversion experiences are gentle and quiet.

Let us pray that we and our loved ones will listen for, and respond to, God's still, small voice—His gentle whisper.

2. The Aldersgate Experience—John Wesley's Conversion

The second conversion story I want to share with you is still subtle, but it happened unexpectedly to John Wesley (1703–91), a theologian of the 1700s. The way he came to accept Christ is called "the Aldersgate experience."

John and his brother, Samuel, both had what John described as "a fair summer religion." Both brothers were ordained, and they both preached, taught, wrote, composed hymns, and even devoted themselves to missionary work. However, they had never accepted Jesus Christ as their Savior. They lived by good works but not by faith.[56]

On the night of May 24, 1738, John Wesley accepted an invitation to a "religious meeting" on Aldersgate Street in London. He went "very unwillingly." During the meeting, while listening to a reading of *Martin Luther's Preface to the Epistle to the Romans*, something supernatural happened. Wesley later wrote that he felt his heart "strangely warmed." He wrote, "I felt I did trust in Christ, Christ alone for salvation; and an assurance was given me that He had taken away my sins, even mine, and saved me from the law of sin and death."[57]

John's brother, Charles, had accepted Christ as his Savior three days earlier. From that point on, John and Charles were passionate about spreading the message of God's mercy and grace. It was that experience that ignited the movement we know today as Methodism. United Methodists throughout the world celebrate Aldersgate Sunday on May 24st, or the Sunday closest to that date, to

---

56 "John Wesley's Reluctance Turned into a Mission After Encounter at Aldersgate," The Florida Conference of the United Methodist Church, https://www.flumc.org/newsdetail/john-wesleys-reluctance-turned-into-an-inferno-after-encounter-at-aldersgate-17424710.
57 Ibid.

remember what happened all those years ago and to celebrate how God, our Creator, can transform people in an instant.[58]

Wesley's experience happened quietly—he felt his heart "strangely warmed." In my experience as a pastor, I have seen this type of conversion pretty often. It is typically a subdued, quiet, and very personal experience—but of course still profound.

3. The Damascus Road Experience—Saul's Sudden, Dramatic Conversion

The third example of a person's conversion is much less subtle. It's the sudden, dramatic conversion called "the "Damascus Road experience" (see Acts, chapter 9).

God, our Creator, knew that to spread the Christian message, He needed a messenger who could speak to both Jews and Gentiles. Saul fit that description, but he was persecuting and killing Christians. Jesus spoke to Saul in a way that only a fool would reject the call!

As Saul was on his way to the city of Damascus to make more arrests, he had a dramatic experience of the risen Jesus. Near Damascus, a bright light from heaven hit Saul; he was knocked to the ground and blinded. Then he heard the voice of Jesus saying, "Saul, Saul, why do you persecute me?" For three days, Saul was unable to see, and he did not eat. Jesus sent Saul to the next town, where he would be told what to do.

Shortly afterward, the Lord spoke to a disciple named Ananias, asking him to go to Saul, place his hands on him, and restore his sight. Ananias was understandably skeptical because Saul was well-known for the harm he did to believers and his plans to arrest more of them. The Lord told Ananias that he had bigger and better plans for Saul, and Ananias obeyed.

---

58 Ibid.

Saul's sight was restored, he became a follower of Jesus, and he changed his name to Paul. And then he became the most visible and active pillar of the new Christian church. He even wrote nearly half of the New Testament. Paul's life changes and contributions are nothing short of a miracle, but it took Ananias and a supernatural, abnormal event to make it happen.

I wonder if somebody was praying for Saul as he went about killing Christians. Perhaps even Christians jailed by Saul prayed something like this:

> Lord Jehovah, I don't know why Saul hates Christians so much. Several of my friends and I are imprisoned because of his persecution. We may even be put to death. That scares me, Lord; still, I pray for Saul, that, by Your Grace, he might come to know Jesus Christ as the true Messiah. May he, by Your grace, embrace Jesus as his own Savior, find Your forgiveness, and embrace Your love. In Jesus's name I pray. Amen.

There is no biblical evidence of this, but maybe some Christians were praying for him to have a change of heart. That would have seemed like an impossible outcome, given how much Saul persecuted Christians. However, nothing is impossible for God.

Saul's "Damascus Road experience" was a dramatic event—much more sensational than the way the Bible usually depicts Jesus's actions. When Saul was knocked to the ground and blinded, Jesus's words to him were harsher than we might expect. Jesus suspended the usual, the expected, to achieve the outcome He desired.

Anyone who knew Saul, and how much he despised Christians, would have never believed that such a transformation could take place. But it did—only through God's power and grace. This is why I continue to pray, every day, that people who aren't following God will have a conversion experience, whether it's subtle or dramatic.

You Never Know when Someone Will Realize They Need Him

There are many examples of people (like John Wesley) who accepted Jesus as their Savior and Lord—surprising themselves as well as the people around them. More recently, Wikipedia cofounder Larry Sanger became an outspoken Christian after spending decades as a skeptic who wasn't sure about where he stood on the topics of God and faith.

In February 2025, Sanger told a reporter at CBN News that he traced the skepticism he held for most of his life back to his childhood and his unsure perspective on all religions. He said he did, however, "dabble in attempts to communicate with a Higher Power." He writes, "I wasn't necessarily praying to God, per se, but I had a sort of internal dialogue. And sometimes, I even wrote it out…with some supremely wise being, and, sometimes, I would even call that being 'God,' not that I believed that that was God, but in order to just sort of clarify my thought."[59]

The process for Sanger was rather subtle and gradual. He said he began thinking more deeply, reading the Bible, and piecing together evidence that what the Bible claims is true. Next, he admitted to himself that he was praying to God, and he prayed the "Sinner's Prayer," which is a common reference to the prayer people use when asking Jesus to come into their hearts. It was a process and takes time because "it involves an entire shift of a worldview."

---

59 "'I Was Wrong': Wikipedia Co-Founder Embraces Bible, Abandons Skepticism," Billy Hallowell, FaithWire, February 28, 2025, https://www.faithwire.com/2025/02/28/i-was-wrong-wikipedia-co-founder-embraces-bible-abandons-skepticism/. See also "How a Skeptical Philosopher Becomes a Christian," Larry Sanger's blog, February 5, 2025, https://larrysanger.org/2025/02/how-a-skeptical-philosopher-becomes-a-christian/.

Finally, he concluded that Scripture is true, validated, and provable. He writes, "The Bible withstands scrutiny, which was a great surprise to me. I thought it couldn't. I was wrong."[60]

So again, please don't give up on praying for your loved ones who have not yet accepted Jesus Christ as their Savior and Lord!

Salvation in the Last Moments of a Person's Life

Continuing the theme of what's common and what's rare, in my experience as a pastor, most people accept Jesus Christ while living their lives—as preteens, teenagers, or adults. I have rarely seen people come to Jesus on their death beds. However, it does happen, and when it does, Jesus is pleased that another one of His loved ones will spend eternal life in heaven with Him.

A parishioner I worked with a long time ago, Dan, told the story of how he was praying for his dad, who was in a coma. His dad had never wanted to accept Christ as his Savior and refused to even believe in God. He was comatose, unable to respond, talk, or move.

Dan became a Christian as an adult. Ever since then, he had been praying for his dad, whom he loved immensely. One day, as Dan sat by his father's side at the hospital bed, he spoke again to his unresponsive father. He told him of Christ's love for him. He told him how Christ died on the cross for him and that all he had to do was call the name of the Lord, and he would be saved.

And then, miraculously, for a brief moment, Dan's dad opened his eyes and said he accepted Jesus Christ. He closed his eyes again and passed away a short time later.

This type of last-minute conversion, which people sometimes call an "eleventh-hour salvation" or "five o'clock salvation" is compelling evidence that we should never stop interceding for

---

60 Ibid.

nonbelievers—our loved ones and even our enemies—to come to know Jesus as their Lord and Savior.

Recent studies show that there is measurable activity in the human brain, even after a person no longer has a heart rate or a pulse.

Dr Stuart Hameroff, an anesthesiologist and professor of anesthesiology and psychology at the University of Arizona, is just one scientist who thinks this end-of-life brain activity may prove that our souls leave our bodies when we die. Hameroff explained in an interview about one patient that, despite the fact that the patient had almost no other signs of life, the electroencephalogram (EEG) reading saw an energy spike in the individual's brain.[61]

In another study, researchers analyzed the EEGs of four dying patients before and after the clinical withdrawal of their ventilatory support. They found stimulated gamma activities in two of the patients. Gamma activity in the brain is high frequency and is associated with cognitive functions like attention, working memory, sensory perception, and information processing. The study authors wrote: "While the mechanisms and physiological significance of these findings remain to be fully explored, these data demonstrate that the dying brain can still be active."[62]

The Final Seven Minutes Before Death

So, why do I continue to pray for loved ones to accept Christ as their Savior, even though they show no interest in doing so thus far? It has to do with what many people call "the last seven minutes of life." This is not a scientifically recognized term, but rather a concept. It refers to the fact that researchers have found that the

---

61 "Brain Activity May Prove Our Souls Leave Our Bodies when We Die," Ellie Abraham, February 25, 2025, MSN, https://www.msn.com/en-ca/health/medical/brain-activity-may-prove-our-souls-leave-our-bodies-when-we-die/ar-AA1zKQYj.
62 Ibid.

human brain continues to work for seven minutes after a person no longer registers a pulse or blood pressure.

Back in chapter 2, I mentioned that Darwin was reportedly estranged from God. However, I wonder if, in the last seven minutes of his life, he might have been restored to a relationship with God. Because his loving wife, Emma, was a believer, I can imagine that she probably shared God's message of love, mercy, and salvation with him—and prayed for him to be saved—until he took his last breath.

In March 2020, doctors in a Canadian intensive care unit found that a patient had sustained brain activity for seven minutes after they turned off his life-support machine. Even after medics declared the person clinically dead, his brain waves continued to occur as if he were sleeping. The researchers also found that the experience of death can vary widely among individual patients.[63]

We are seeing more and more studies suggesting that the human brain is capable of well-organized electrical activity in the early stages of clinical death. In fact, in some studies, the low gamma waves produced during the early stages of consciousness, such as in the first few minutes after death, became stronger during this short period of time. Canadian doctors in intensive care seem to have observed that a person's brain continues to function even after clinical death. One researcher wrote, "This suggests that our final journey into permanent unconsciousness may indeed involve a short state of heightened consciousness and memory before we die."[64]

The human soul does not reside in the cardiac muscle; in my opinion, it resides in the brain. That's why during the final seven minutes of a person's life, a lot of repenting and confessing can happen. A person who has rejected Christ for his or her entire life—even for decades—will get one final chance to reconsider their long-

---

63 "7 Minutes After Death: You're Alive," Medium, July 30, 2020, https://medium.com/sciention/7-minutes-after-death-youre-alive-8a407d42e32.
64 Ibid.

held disbelief and to finally listen to what many loved ones may have told them: that they can live eternally with God and with their loved ones if they will only accept Jesus Christ as their Savior.

Never stop praying for nonbelievers, including your loved ones. Who knows? During those final minutes of life, they could have a Damascus Road experience, an Aldersgate experience, or a small, still voice experience. They might remember every word of the testimony you've shared with them—whether just once or many times over the years—and your prayers might finally be answered. Pray for them to know that God loves them and wants them to have a personal relationship with Him.

The Parable of the Vineyard Workers

God bestows the gift of eternal life in heaven with Him on everyone who trusts in Him and accepts Him as their Savior.

Whether someone accepts Christ into their heart in the last seven minutes of life or was saved as a child and lives to be 100 years old, the blessing is the same. Just as it doesn't matter how subtle or dramatic your conversion is, it doesn't matter when you accept Him. Of course accepting Him into your heart at a young age enables you to serve Him for your entire life, which pleases Him. But if you become a Christian as you're taking your last breath of life, you will still reside in heaven with Him forever.

Jesus demonstrated this concept in "the parable of the vineyard" in Matthew 20:1–16.

In that Scripture, Jesus tells a parable about vineyard workers going out to work. Some went early in the morning, some at mid-morning, some at noon, some at mid-afternoon, and some with just an hour left of work.

When it came time for them to get paid, the master of the house gave the same pay to those who worked just one hour to those who had worked all day in the scorching heat. Those who had worked all day were angry with the master. His reply to them showed that he valued their work equally, regardless of what time they began working:

> "Take what belongs to you and go. I choose to give to this last worker as I give to you. Am I not allowed to do what I choose with what belongs to me? Or do you begrudge my generosity?"
>
> **—Matthew 20:14–16**

I find great comfort in this wonderful truth about God's generosity, salvation, and eternal life.

Knowing that God loves all of us equally—no matter when someone accepts Him as their Savior—helps me continue to pray fervently for my loved ones who are not believers. It should remind all of us that not only is He magnificently powerful; He is also the extremely loving God who wants us to live with Him eternally. We are called to honor and glorify Him while we are living on Earth. He will not give fewer blessings to those who come to Him in repentance in the eleventh hour.

As you pray for your loved ones, I recommend holding a photo of them. This can make your prayer experience more personal and powerful.

A Prayer to Our Heavenly, Holy Creator, That Your Grace Might Warm the Hearts of My Unbelieving Family and Friends to Embrace Your Truth

Heavenly Father, I come to You embracing the truth that You are all powerful and all-knowing. You created everything there is, was, and ever will be on this Earth and in this universe.

My heart today is somewhat heavy for some of my friends and family who do not know you as their Lord and Savior. Some of them, Lord, have outright rejected you. They have embraced the lies and hate of the evil one rather than the truth and love Your grace provides.

Sometimes, Father, when I reflect back on my life and on the lives of those I love, I see how my life negatively impacted them in their faith. I haven't always lived my life with Your love as my encouragement and with the truth of Your Word as my guide. Forgive me, Lord, for these lapses in judgment and the sins I have committed. Heal my loved ones of the wounds I have caused them, both intentionally and unintentionally.

I pray that my loved ones do not wait till the eleventh hour to give their lives to You, to embrace You as their Savior and Lord, because I know their lives would be more fulfilling if they started following You today. In my limited human view, Lord, I think it's going take a Damascus Road experience to knock some of them out of their unbelief and into faith. I acknowledge that such a dramatic type of conversion is not Your normal way of doing things. More often, people come to you through more of the Aldersgate experience of John Wesley or the small, still voice experience of Elijah.

My prayer, Lord, is that by whatever means, and on whatever timeline, is necessary, my loved ones will come to know You as their Savior first and then as their Lord. Whether it happens today or in the last moments of their lives, I give you praise. You created our brains with the ability to do marvelous things, even in the last moments of life.

In Jesus's name I pray. Amen.

Actions I Will Take:

1. Continue to be faithful in prayer for those who seem distant from Christ.
2. Try to be an example of Christ's love to them.

Now I encourage you to write your own prayer. Think about what we have discussed in this chapter, and pour out your heart to Him. What do you want to say to Him? What do you want Him to communicate to you?

____

____

____

____

____

____

____

Actions You Will Take:

1. ________________________________
2. ________________________________
3. ________________________________

CHAPTER 10
PRAYING AS A CATALYST TO HEAL RELATIONSHIPS

Just like a catalyst facilitates metabolic processes, our relationships need catalysts to move forward. Prayer and open communication are two catalysts that enable relationships to thrive.

When we think of a *catalyst*, we often think of an agent that facilitates chemical/biochemical processes, such as metabolic pathways. Here is one definition, from Merriam-Webster, that relates to such processes: " A *catalyst* is a substance that enables a chemical reaction to proceed at a usually faster rate or under different conditions (as at a lower temperature) than otherwise possible."

We have catalysts in everyday life, as well. A second definition from Merriam-Webster applies to catalysts of the non-scientific type: "An agent that provokes or speeds significant change or action."

For example, the *Apollo 8*'s "Earthrise" photo shown in chapter 6 served as a catalyst that helped launch a worldwide environmental movement aimed at protecting the Earth for future generations.

"Earthrise," the first color image of the Earth taken from outer space, enabled people to see, for the first time, what our home planet looks like from 238,855 miles away. Whereas our usual perspective is that the Earth is vast, expansive, and blessed with what seems like infinite resources, this new perspective made people realize that the Earth is actually delicate and vulnerable.[65]

The photo was a catalyst for the founding of the modern environmental movement. A little more than one year after "Earthrise" was distributed, the world celebrated its first Earth Day. By the end of the year, the US government created the Environmental Protection Agency and passed trailblazing environmental legislation like the Clean Air Act, the Clean Water Act, and the National Environmental Education Act.[66]

A person, event, or thing can be a catalyst to bring about change or action.

We also have catalysts in our relationships. Like scientific processes, our relationships with others are complicated and often need some sort of catalyst to help them move forward—and, if necessary, to heal. Prayer, communication, and gestures of kindness are just some of the catalysts that can help our relationships thrive.

When I talk with people about relationship issues, sometimes they make the comment, "It's complicated." That's true of any relationship, whether it's with our parents, siblings, children, friends, colleagues, supervisors, or a spouse. In-laws are sometimes outlaws, and siblings can be enemies.

When relationships are broken, it can seem like nothing will ever heal them. Yet God can!

---

65 "Climate Action: The Photo that Captured the World," Earth Day website, June 13, 2024, https://www.earthday.org/the-photo-that-captured-the-world/.
66 Ibid.

All of God's creations are complicated. Metabolic pathways, stars and galaxies, the sophisticated functions of neurotransmitters, the beauty and complexity of music—everything discussed in this book and beyond—are examples of intricate, complex processes that God created. Yet they are all part of His magnificent creation, and they function in perfect harmony with the rest of the universe. He can restore them when they malfunction—just as He can restore damaged relationships.

How Enzymes Convert Glucose to ATP

A metabolic process has to take place during *aerobic respiration* for enzymes to convert glucose (a type of sugar) to adenosine triphosphate (ATP), the primary energy-carrying molecule in cells. Biologists often call ATP the "energy currency of the cell" because it is used to power various cellular processes.

Glycolysis is the first stage. This is the process by which glucose is broken down into smaller molecules called pyruvates. A *pyruvate* is a critical three-carbon compound that can be used for various purposes, including energy production, *gluconeogenesis* (a process in which the body generates glucose, or sugar, from non-carbohydrate sources, primarily in the liver and kidneys), and the synthesis of amino acids and fatty acids. This process occurs in the *cytoplasm* (the fluid inside the cell) of nearly all living organisms. It is a crucial step in cellular respiration, the process that cells use to extract energy from food.

Next comes the Krebs cycle, in which the pyruvate molecules are oxidized to generate ATP, which is then used in the electron transport chain. This process takes place in the *mitochondria*, considered the powerhouses of a cell because they are responsible for generating most of the cell's energy.

The following is a somewhat simplified visual depiction of the chemical process involved in aerobic respiration:

Aerobic Respiration

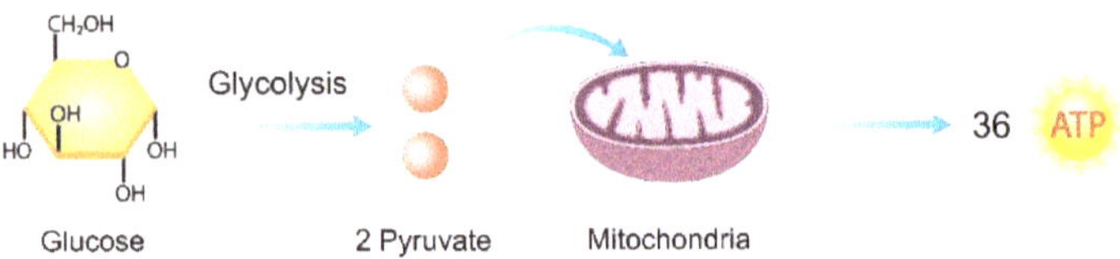

For the Krebs cycle to work, eight major enzymes are required to catalyze the reactions that convert one molecule into the precursor for the next molecule in the cycle. It's like taking a "C" and adding a line above the bottom curve so it becomes a "G."

We need ATP for our muscles to contract; for our brains to transmit neurosignals; and for the construction of DNA, RNA, proteins, lipids, and other tiny but complex mechanisms in our bodies. Without ATP, we would not be able to move our muscles or even have thoughts.

The purpose of this very complicated biological reaction is to create energy that enables cells to do their work within our bodies.

This is a fascinating but complicated process. We can really see the sophisticated design of this process when we ask, "Which one of these enzymes came first?" As I attempt to answer this question, remember that in chapter 1, I noted that, according to the theory of evolution, any particle that is useless—serves no particular purpose—will eventually disappear because it becomes a burden for an organism's cells to carry.

In the metabolic process, if a muscle cell appears first, but without ATP, that cell is totally useless. According to the theory of evolution, if the muscle appears first, but without ATP, the muscle will disappear eventually. However, if the first enzyme in the Krebs cycle appears at the same time as the muscle, but there is still no

ATP because the other seven enzymes still have to come around, the muscle will be of little or no use.

If a mutation occurs for the first enzyme to be produced, but there's no mutation to cause the second enzyme to be produced, then the first enzyme is absolutely useless. And if you don't have the mitochondria where all these processes happen, the Krebs cycle cannot occur at all.

It's kind of a "chicken and egg" question.

Later in chapter 1, I mentioned a researcher named Michael Behe, PhD, the author of *Darwin's Black Box: The Biochemical Challenge to Evolution.* As stated, the biochemist argues that "the existence of many complex systems rules out evolution because they are 'irreducibly complex.' That means they cannot function if they are missing just one of their many parts."

Metabolic pathways are "irreducibly complex" systems. They cannot function unless *all* the elements are present. Behe says this is evidence of "intelligent design."

If you find this discussion difficult to understand, you are not alone! It took researchers more than one hundred years to understand the Krebs cycle as the central metabolic pathway.

It all began when the pioneering Swedish–German chemist Carl Wilhelm Scheele isolated citric acid from lemon juice in 1784. Many other scientists also contributed to the discovery and establishment of the citric-acid cycle. Two key researchers were Albert Szent-Györgyi at the University of Szeged (Hungary) and Hans Adolf Krebs at the University of Sheffield (UK). They were awarded the Nobel Prize in Physiology or Medicine in 1937 and 1953, respectively.[67]

---

67 "Citric Acid," American Chemical Society, April 4, 2022, https://www.acs.org/molecule-of-the-week/archive/c/citric-acid.html.

I believe God created metabolic pathways when He said, in Gensis 1:3, "Let there be light." I am in awe of how He created such magnificently complicated systems out of nothing.

The One Who Created Everything Can Also Heal Our Family Relationships

Some of the most difficult experiences we encounter as fallen humans involve broken relationships with family members. I have experienced such frustrations and disappointments in my own life, beginning in childhood.

When I was just two years old, my biological father left my family. I saw him briefly only once after that. Thankfully, my mother later married a really good man, and he provided well for our family. He was a wonderful husband to my mom and a dad to my sisters and me. He was a positive male role model. In fact, my son legally changed his last name to that of my stepfather because he has such great respect for him—and not much respect for my biological father.

More recently, a long-term rift erupted in my family again, causing a lot of stress for all of us. As is the case for many families, our dynamics are complicated. Not only can God heal these relationships; He wants us to repair them and attempt to live in harmony with one another, as much as we are able.

"May the God of endurance and encouragement grant you to live in such harmony with one another, in accord with Christ Jesus, that together you may with one voice glorify the God and Father of our Lord Jesus Christ."

—Romans 15:5–6

For me personally, witnessing how masterfully God has created all the intricate and complex systems in existence today gives me assurance that He can heal relationships. I know that if God can produce all these intricate systems in our bodies and in the universe, He absolutely can heal our complicated relationships.

There is no reason to lose hope, no matter how strained our relationships might be.

In our families, the "chicken-and-egg" or "Which comes first?" conundrum is that, for our relationships to be restored, one individual needs to make the effort to heal divisions. If you have not experienced this type of complicated family dynamics personally, you've probably seen it surface in other families you know. People often say, "I won't apologize until he/she apologizes." If that's the case, no breakthroughs or healing will ever occur. It is only through the divine intervention and influence of our heavenly Father, our Creator, that relationships can heal.

Catalysts are important elements of metabolic processes and family dynamics alike. Prayer can be the catalyst that helps repair relationships. As Christians, let us have the humility and courage to reach out to our loved ones and try to restore harmony. Let us use prayer as the catalyst that stops the discord.

However, I believe not every relationship is meant to be restored. Sometimes, it is God's will for someone to *leave* a toxic and/or abusive relationship.

When I was a pastor and advised couples, there were times when I advised a spouse who was experiencing abuse in the relationship to leave. I gave such advice only after much reflection and prayer. Everyone needs to feel, and be, safe in a relationship.

It is never good for anyone to stay in a toxic or abusive relationship. We can always pray from a distance, but not all relationships can be restored. You are a beloved child of God, and He does not want His children abused in any way, whether it's sexual, physical, verbal, or emotional in nature. These dynamics are not of God, and they typically render a relationship irreparable.

Seek guidance, prayer, and help from your church or from family members and friends. Pray about the situation. That cosmic communication with your Creator can help you discern when and how to escape such a situation.

"And a harvest of righteousness is sown in peace by those who make peace."

—James 3:18

Never lose hope! Take inspiration from our Creator's wonderful workmanship, and put all your trust in Him. Pray to him unceasingly—about *everything*.

A Prayer to Our Holy Father, to Heal Broken Relationships

Holy and Heavenly Father, Creator of all the biochemical processes that keep my body going, thank You for enabling my brain to process complex thoughts and feelings like love and concern. You know how complicated my family relationships are because you knew me before I came to exist. And You know the impact those unhealthy dynamics have had on generations of my family. I do not know how to heal the brokenness.

Thank You for bringing my stepfather to our family. He was so good to my mother, my sisters, and me, and he was a great role model for all of us, as well as for my son and daughter. That shows your ability to heal complicated family dynamics, just as your creation of complicated metabolic pathways demonstrates your ability to create, maintain, and restore the functioning of all living organisms.

Thank You for the healing that has begun in my family. I pray that the healing continues until all the complications disappear. I realize that if this restoration is Your will, it could happen during my lifetime or later.

I submit myself to Your will. Give me the strength, courage, and wisdom to seek peace and to be a part of this healing process. In Jesus's name I pray. Amen.

Actions I Will Take:

1. Be sensitive enough to notice when someone might be in an abusive relationship.
2. Pray, offer encouraging words and, if appropriate, help someone escape from an abusive relationship.

Now I encourage you to write your own prayer. Think about what we have discussed in this chapter, and pour out your heart to Him. What do you want to say to Him? What do you want Him to communicate to you?

____

____

____

____

____

____

____

____

____

____

____

Actions You Will Take:

1. ______________________________________
2. ______________________________________
3. ______________________________________

CHAPTER 11
BEGINNING COSMIC CONVERSATIONS WITH OUR CREATOR

Throughout this book, I have compared and contrasted scientific facts and theological truths in an attempt to encourage all of us to be in awe of God as our Creator and to pray to Him fervently—about everything.

It takes time and repeated effort to break bad habits and to develop healthy new habits. Pray all the time to develop this habit that is so important to our spiritual and overall well-being!

Four Tips for Developing the Habit of Praying

Here are four practical ways to begin developing the prayer habit.

1. **Notice everything around you.** No matter how big or small, simple or complex something is, acknowledge that God created it. Appreciate it, and thank Him for His creations. When something happens that catches your eye, pray for the people in that situation. God brings our attention to whatever, or whomever, He wants us to focus on.
2. **Turn to Him in times of distress.** When you are experiencing a difficult situation, pray for insight, wisdom, and comfort. Know that all things work together for His glory. Ask Him to show you what He wants you to learn from those experiences.
3. **Ask God what He wants you to do next.** Prayer is a powerful act that lifts others up and brings us closer to

Him. However, often prayer is the first step in God's plan for what He wants us to do. After you pray, ask God, "What do You want me to do next?"

For example, if you pray for a homeless person you see on the street, ask God what He wants you to do next to help homeless people. Maybe you could volunteer at a shelter or food pantry. Chances are, if something tugs at your heart, God is speaking to you!

4. **Notice your emotions.** Pay attention to how you're feeling. Use the Scriptures and sample prayers that follow to ask for His help in navigating situations that cause these emotions to surface in your daily life.

Scriptures, Sample Prayers, and Action Items to Help Navigate Everyday Emotions, Struggles, and Situations

In this section are sample prayers you can modify as you pray to God while navigating everyday emotions, struggles, and situations. They appear in alphabetical order—except for the first prayer for salvation, because that is the most important prayer of all.

What I have found helpful when faced with a situation is to find a passage of Scripture that addresses that situation, or one like it. For that reason, I have included a relevant passage of Scripture before each of the sample prayers in this chapter to provide you with inspiration from God's Word.

I encourage you to use these prayers as a starting point. Modify them as needed so they relate to your situation. Pray to Him in detail, about everything—from the depths of your heart. Give Him praise for the blessings He has already given you. Pray not only for your own situation but for others as well—be an intercessor.

So here they are—Scriptures and sample prayers for some of the many turning points in life.

Salvation

I have placed this prayer for salvation first because it is the most important prayer any individual can pray to God. It is the prayer you pray to ask Jesus to come into your heart, to become your Personal Savior. When you pray this prayer, you can begin your relationship with our Creator and have cosmic conversations with Him.

> "For God so loved the world, that he gave his only begotten Son, that whosoever believeth in him should not perish, but have everlasting life."
>
> **—John 3:16, KJV**

Jesus, right now, I ask you to become Lord of my life and my Savior, and I thank You for the sacrifice You made on the cross. I receive Your grace and confidently believe that my life is forever changed.

Jesus, help me to live for You. My heart and my life are open to what You have for me. I want to know Your ways and love people as You love me. I submit everything within me to You. Thank You for Your mercy, hope, and unfailing love.

In Your name I pray. Amen.

Addiction

"Brothers, if anyone is caught in any transgression, you who are spiritual should restore him in a spirit of gentleness. Keep watch on yourself, lest you too be tempted."

—Galatians 6:1

Heavenly Holy Father, Creator, all that there is, You created us in Your image, and what a beautiful image that is! Yet we are fallen and often struggle with substance and/or behavioral addictions. I pray for [*person's name*], who is struggling to escape the clutches of [*name the specific addiction, such as alcohol, drugs, gambling, pornography, etc.*]. Let me be a good friend to her. Empower me to help her get the support she needs, whether through Alcoholics Anonymous/Narcotics Anonymous or in-patient care— whatever it takes. May I have the willingness to bear her burdens throughout this struggle. May I do so with the gentleness of Your Son, Jesus Christ, without judgment.

In Jesus's name I pray. Amen.

Anger

"Be angry and do not sin; do not let the sun go down on your anger."

—Ephesians 4:26

Heavenly Father, in the name of Your Son, Jesus Christ, I ask that you forgive me for the times when I've sinned by expressing anger that is a result of bitterness and revenge. I ask that You encourage me and empower me to make amends for the times when my sinful anger has harmed others or shaken their faith in You because of my actions. I know that in the future, there's a good chance that anger may arise within my heart again. When it does, Lord, replace my anger with discernment and restraint.

In Jesus's name I pray. Amen.

Anxiety

"Do not be anxious about anything, but in everything by prayer and supplication with thanksgiving let your requests be made known to God. And the peace of God, which surpasses all understanding, will guard your hearts and your minds in Christ Jesus."

—Philippians 4:6–7

Lord Jesus, there are so many things in this world that impact my anxiety. I am often anxious about my finances, health, and relationships. I pray that You will help me remember the Scripture above when I become anxious. Empower me to push worry aside, go to You in prayer, and trust You to bring about peace that surpasses all understanding.

In Jesus's name I pray. Amen.

"If possible, so far as it depends on you,
live peaceably with all."

—Romans 12:18

Lord, as I drive through downtown Denver today, I am reminded that we, as Christians, need to see the needs of the individuals in our city on every single block.

I see men and women walking and driving. They represent all different ages, nationalities, languages, and financial situations. I see apparently wealthy people wearing designer outfits, as well as homeless folks on the street, asking for help. I see people working in construction and others working in hospitality. I see students on campus, businesspeople going to lunch, medical personnel taking a break, first responders racing to emergencies with their sirens on, and law enforcement officers pursuing suspects.

Every single one of these people, Lord, has a need for a deeper relationship with You. Some of them know You intimately and have committed their lives to You following a powerful plea for pardon. Others have rejected You outright because they've been hurt by the church or for some other personal reason.

Father, help me notice those who are hurting, have compassion toward them, and help them in a meaningful way that will help lift them out of their despair. Help turn my compassion into action. Do you want me to serve in a homeless shelter? To volunteer at a treatment center? Show me how I can help these people beyond praying for them.

Lord, use the skills, talents, and gifts of service You have given me to empower people to live lives of faith, love, and grace through compassion, kindness, and truth. Father, sometimes the truth is unpopular.

Lord, I thank You for showing me the deep needs of humanity located in this city and for leading me to realize that I am not always as compassionate as I should be. May this not just be a prayer asking You to help these folks, but a prayer for You to show me, tell me, and push me to do whatever I can to help people beyond praying for them—and without enabling them to be dependent on others.

I'm reminded of Matthew 5:40 in Your Word: "And if any man will sue thee at the law, and take away thy coat, let him have thy cloak also." Lord, I am at a loss about how to be compassionate without enabling people, about how to be kind without being foolish.

Show me what You want me to do, Lord.

In Jesus's name I pray. Amen.

Decision Making

"If any of you lacks wisdom, let him ask God, who gives generously to all without reproach, and it will be given him."

—James 1:5

Today, Lord, I pray for a friend of mine who is very close to my heart. His father is very ill at this time, and my friend is trying to decide whether he should take a mission trip outside the country this month or stay home and be with his elderly and ailing dad. Grant wisdom and insight to him, his wife, his children, and his extended family so they make the right decision. Encourage them all to seek Your wisdom as they decide what to do.

In Jesus's name I pray. Amen.

Discouragement

"I have said these things to you, that in me you may have peace. In the world you will have tribulation. But take heart; I have overcome the world."

—John 16:33

Lord Jesus Christ, You told us that we will have tribulations in this world that can lead to discouragement and disappointment. I pray, Lord, that when I am discouraged, I will remember the verse above, in which you promise that You have overcome this world, which is only temporary for Your followers. May I live in the peace of knowing that You are in control.

In Jesus's name I pray. Amen.

Dislike

"Bless those who persecute you; bless and do not curse them. Rejoice with those who rejoice, weep with those who weep. Live in harmony with one another. Do not be haughty, but associate with the lowly. Never be wise in your own sight. Repay no one evil for evil, but give thought to do what is honorable in the sight of all. If possible, so far as it depends on you, live peaceably with all. Beloved, never avenge yourselves, but leave it to the wrath of God, for it is written, 'Vengeance is mine, I will repay, says the Lord.' To the contrary, 'if your enemy is hungry, feed him; if he is thirsty, give him something to drink; for by so doing you will heap burning coals on his head.' Do not be overcome by evil, but overcome evil with good."

—Romans 12:14–21

Dear Father in heaven, please help me love [*person's name*] the way You love him/her and the way You love me. Please remove the animosity I feel toward this person.

In Jesus's name I pray. Amen.

Envy

"Love is patient and kind; love does not envy or boast; it is not arrogant."

—1 Corinthians 13:4

Heavenly Father, I know that throughout my life, I have often found myself to be envious of people who seem to have a better life than I do. I have felt envious about money, relationships, and, more recently, health. Lord, forgive me for these feelings of envy and jealousy. I know that the Scripture above is from what is known as "the love chapter" of the Bible. Chances are, when I am envious or jealous of somebody, it's because I don't love them enough. Help me to love deeply and to be thankful for the life You have given to them and to me.

In Jesus's name I pray. Amen.

Fear

"Fear not, for I am with you; be not dismayed, for I am your God; I will strengthen you, I will help you, I will uphold you with my righteous right hand."

—Isaiah 41:10

In Jesus's name, I come before You, Lord, knowing full well that a lot of things in this life are frightening. This includes wars, rumors of wars, violence around the world, school shootings in our own neighborhoods, and many other things. I pray for the people in war-torn areas of the world, that their fear will not turn to dismay. May they turn to You and seek Your strength, presence, and comfort, even in the midst of fear.

In Jesus's name I pray. Amen.

Forgiving Myself

"If we confess our sins, he is faithful and just to forgive us our sins and to cleanse us from all unrighteousness."

—1 John 1:9

Jehovah Jireh, I confess to you that sometimes I have trouble forgiving myself, especially for sins that seem to be repetitive. When I am trying to forgive myself, please bring the above verse to my mind, Lord. Empower me to confess my sins to You. Remind me of Your faithfulness. Lead me to embrace Your forgiveness.

In Jesus's name I pray. Amen.

Forgiving Others

"Put on then, as God's chosen ones, holy and beloved, compassionate hearts, kindness, humility, meekness, and patience, bearing with one another and, if one has a complaint against another, forgiving each other; as the Lord has forgiven you, so you also must forgive."

—Colossians 3:12–13

Lord Jesus Christ, whenever I am struggling to forgive someone, I pray that You will help me have a compassionate heart filled with kindness, humility, and patience. Remind me how much, and how many times, You have forgiven me. Let Your grace guide me to forgive those who have harmed me.

In Jesus's name I pray. Amen.

Enemies

"But I say to you who hear, love your enemies, do good to those who hate you, bless those who curse you, pray for those who abuse you."

—Luke 6:27

Heavenly Father, this may be one of the most difficult passages of Scripture for me. No one really wants to do good to those who hate us. It's hard to bless those who curse me. It's hard to pray for, and forgive, those who have abused me, Father, but that's what Your Son did while He walked on this planet. Empower me, by Your grace, to live, act, speak, and love as Jesus did.

In Jesus's name I pray. Amen.

Evil

"For we do not wrestle against flesh and blood, but against the rulers, against the authorities, against the cosmic powers over this present darkness, against the spiritual forces of evil in the heavenly places."

—Ephesians 6:12

Lord Jesus Christ, there are so many acts of violence and other forms of evil in this world. Human trafficking, murder, lies, greed, road rage, crime, and sexual perversion are just a few. Whenever I hear of such things, Lord, remind me of Ephesians 6:12.

Let me hate acts that are evil, Lord, but to love everyone, even if they do something that is wrong in Your eyes. You loved me when I walked in darkness. You even loved and forgave those who hung You on the cross. By Your grace, empower me to love like You. Also help me to put that love into action whenever I interact with people who may simply disagree with me.

In Jesus's name I pray. Amen.

"And hope does not put us to shame, because God's love has been poured into our hearts through the Holy Spirit who has been given to us."

—Romans 5:5

Lord Jesus Christ, often I feel guilt or shame about things I have done—from the distant and more recent past. Whenever that happens, please help me remember that the Holy Spirit has poured Your love into my heart. May I embrace it. I am grateful that, through the death and resurrection of Jesus Christ, my relationship with You has been restored. May Your grace and forgiveness prevent me from wallowing in shame, guilt, or regret. May I also comfort others with the truth of the Scripture above.

In Jesus's name I pray. Amen.

"Trust in the Lord with all your heart, and do not lean on your own understanding. In all your ways acknowledge him, and he will make straight your paths."

—Proverbs 3:5–6

Lord God Almighty, thank You for sending Your only Son to die on the cross so all my sins are forgiven. I pray that in all the decisions I make in any area of my life, I will trust You completely and totally to help me make decisions based on that trust. May I acknowledge You in every decision-making process, and may Your grace, love, and hope guide those decisions.

In Jesus's name I pray. Amen.

Resentment

"Let all bitterness and wrath and anger and clamor and slander be put away from you, along with all malice."

—Ephesians 4:31

Heavenly Holy Father, in Jesus's name, I ask that, whenever I feel resentful, may You reveal the bitterness, wrath, and anger that might be within my heart. Remind me that You have forgiven me, empowered me by Your grace, and set me free in the hope of the Cross.

By Your grace, remove all resentment from my heart.

In Jesus's name I pray. Amen.

Revival

"If my people who are called by my name humble themselves, and pray and seek my face and turn from their wicked ways, then I will hear from heaven and will forgive their sin and heal their land."

—2 Chronicles 7:14

Lord Jesus, as I read and hear about all the troubles that seem to infect the United States of America, all of them seem to be a result of our wandering away from Your truth, grace, love, and hope. We often push faith aside and embrace the lies of the evil one. I come before You, Father, knowing full well I have fallen short of Your glory. I seek Your face in all things. May we, as a nation, turn from our wicked ways. Let us pray to you, Lord, seek You, and embrace Your ways. I pray that You hear us and that You will heal this land and our hearts. Remove the divisions, and stop the lies.

In Jesus's name I pray. Amen.

The Seven F's

"Do not be anxious about anything, but in everything by prayer and supplication with thanksgiving let your requests be made known to God."

—Philippians 4:6

Lord God, I pray that my family, my friends, my fellowship, and my fiefdom will be filled with faith. Help me embrace the future with forgiveness.

In Jesus's name I pray. Amen.

Temptation

"And when he came to the place, he said to them, "Pray that you may not enter into temptation..." "And he said to them, 'Why are you sleeping? Rise and pray that you may not enter into temptation.'"

—Luke 22:40, 46

Lord, in the garden right before Your arrest, You told the disciples to pray that they might not enter into temptation. You knew full well that heartbreaking events were soon to come. You knew how tempting it would be for the disciples to succumb to despair and doubt and to be consumed by anger. Lord Jesus, it seems as if every day, we face heartbreaking events. When this happens, I pray that I will not be tempted to give up, run away, or allow anger and doubt to consumer me. Empower all your followers, including me, to stand firm in faith, no matter what comes our way.

In Jesus's name I pray. Amen.

Wisdom

"Let the word of Christ dwell in you richly, teaching and admonishing one another in all wisdom, singing psalms and hymns and spiritual songs, with thankfulness in your hearts to God."

—Colossians 3:16

El Shaddai, we live in a complicated world. It is even more complicated because of the lies the evil one tells. May I worship You and read Your Word on a regular basis so Your wisdom is in my heart and mind.

In Jesus's name I pray. Amen.

World Conflicts

"You will hear of wars and rumor of wars. See that you are not alarmed, for this must take place, but the end is not yet. For nation will rise against nation, and kingdom against kingdom, and there will be famine and earthquakes in various places. All these are at the beginning of the birth pains."

—Matthew 24:6–8

Heavenly Holy Father, as I read the Scripture above, I think of wars and events that are happening in the Middle East, Ukraine, and many other places around the globe. It is heartbreaking that there is so much hatred and animosity among humanity created in your image. Truly, the evil one is still seeking to destroy and devour. And at times he is succeeding. His lies, when believed, bring about hate, death, and destruction. Frustrate and reveal his lies.

Oh Lord, bring about healing peace, unity, forgiveness, and reconciliation. Be with those in the conflict zones. Lord. Give comfort where there is loss, healing where there are wounds, and encouragement where there is despair.

Protect the innocent, Father. Provide strength and courage to those who are walking in the light, and frustrate the plans of those who are serving the darkness. Strengthen the faith of believers in zones of conflict, as well as the faith of believers in the midst of natural disasters. May faith, hope, and love overcome despair and destruction.

In Jesus's name I pray. Amen.

As I finished writing this book, the well-known hymn "How Great Thou Art" kept coming into my mind over and over again. I cannot reprint the lyrics here because of copyright issues; however, I encourage you to listen to the song or read the lyrics.

To me, this song beautifully conveys how magnificent and powerful our loving God is. When we recognize Him as our Creator of everything, great and small, we can't help but praise Him and proclaim how great He is!

ABOUT THE AUTHOR

Eugene "Chip" Aydelott, M.Div.

Littleton, Colorado

(303) 988-4875

chip@cosmicconvos.net

Eugene "Chip" Aydelott grew up on an alfalfa and sunflower farm in Fort Sumner, New Mexico, a small town in the eastern part of the state. Chip's dad was a farmer; previously, he was in the US Air Force and served in Vietnam. He was a crew chief on B-52 and B-58 bomber aircraft. For many years, he worked with the Strategic Air Command at Edwards Air Force Base in California.

Chip's mother was a lifelong educator who completed both her bachelor's and master's degrees in education. She taught numerous grades in elementary school and ended her career as a kindergarten teacher.

One day when he was in elementary school at Edwards AFB, Chip's worst nightmare came true. His teacher was absent that day, and he realized that his mother was the substitute teacher!

During high school, Chip became a Christian at the United Methodist Church in Fort Sumner.

After graduating from high school, he attended New Mexico State University in Las Cruces, where he graduated with a bachelor's degree in biology, with an emphasis on human physiology. While in college, he served as the Youth Pastor at University Methodist

Church. He also worked in the intensive care unit at a hospital for two years.

Following graduation from NMSU, Chip wanted to explore missionary work but did not feel ready to become a missionary. He joined the Peace Corps in Costa Rica. For two and a half years, he specialized in aquaculture science—fish farming. He also focused on developmental work, recommending projects related to bridge building and school building and then evaluating the completed projects.

While in Costa Rica, Chip felt God calling him into the ministry. A missionary he knew there recommended that he attend Asbury Theological Seminary in Wilmore, Kentucky. Chip enrolled at Asbury and completed his M.Div. degree. Then he accepted an assignment with the Wesleyan Church to pastor a small neighborhood church in Denver, Colorado, where he remained for twenty years.

He has two adult children. His son, Rob, served in the US Marine Corps and works in IT. He is married to his high school sweetheart, Taylor, who is an intellectual property attorney. Taylor considers Chip to be her dad, and he considers her to be his daughter. Chip's daughter, Alexis, married in 2024 and works in people partnering. Her husband, Aaron, works in a geotechnical laboratory. Chip is extremely proud of his children and their spouses; they are all successful in their respective professional fields.

To supplement his income while he was a pastor, Chip opened a medical supply store in 2006, which he still owns and manages today.

In his spare time, Chip enjoys astrophotography, playing his violin, riding motorcycles, flying radio-controlled airplanes, fly fishing, and, of course, playing pickleball.

www.ingramcontent.com/pod-product-compliance
Lightning Source LLC
LaVergne TN
LVHW010903110826
845149LV00005B/1452

9798994821701